Dr. Chandan Deep Singh
Rajdeep Singh
Ripandeep Singh
Dr. Kanwaljit Singh

Effect of tool pin profile on microstructure and mechanical properties of AL6063 in Friction stir processing

Anchor Academic Publishing

Singh, Chandan Deep, Singh, Rajdeep, Singh, Ripandeep, Singh, Kanwaljit: Effect of tool pin profile on microstructure and mechanical properties of AL6063 in Friction stir processing, Hamburg, Anchor Academic Publishing 2017

Buch-ISBN: 978-3-96067-205-0
PDF-eBook-ISBN: 978-3-96067-705-5
Druck/Herstellung: Anchor Academic Publishing, Hamburg, 2017

Bibliografische Information der Deutschen Nationalbibliothek:
Die Deutsche Nationalbibliothek verzeichnet diese Publikation in der Deutschen Nationalbibliografie; detaillierte bibliografische Daten sind im Internet über http://dnb.d-nb.de abrufbar.

Bibliographical Information of the German National Library:
The German National Library lists this publication in the German National Bibliography. Detailed bibliographic data can be found at: http://dnb.d-nb.de

All rights reserved. This publication may not be reproduced, stored in a retrieval system or transmitted, in any form or by any means, electronic, mechanical, photocopying, recording or otherwise, without the prior permission of the publishers.

Das Werk einschließlich aller seiner Teile ist urheberrechtlich geschützt. Jede Verwertung außerhalb der Grenzen des Urheberrechtsgesetzes ist ohne Zustimmung des Verlages unzulässig und strafbar. Dies gilt insbesondere für Vervielfältigungen, Übersetzungen, Mikroverfilmungen und die Einspeicherung und Bearbeitung in elektronischen Systemen.

Die Wiedergabe von Gebrauchsnamen, Handelsnamen, Warenbezeichnungen usw. in diesem Werk berechtigt auch ohne besondere Kennzeichnung nicht zu der Annahme, dass solche Namen im Sinne der Warenzeichen- und Markenschutz-Gesetzgebung als frei zu betrachten wären und daher von jedermann benutzt werden dürften.

Die Informationen in diesem Werk wurden mit Sorgfalt erarbeitet. Dennoch können Fehler nicht vollständig ausgeschlossen werden und die Diplomica Verlag GmbH, die Autoren oder Übersetzer übernehmen keine juristische Verantwortung oder irgendeine Haftung für evtl. verbliebene fehlerhafte Angaben und deren Folgen.

Alle Rechte vorbehalten

© Anchor Academic Publishing, Imprint der Diplomica Verlag GmbH
Hermannstal 119k, 22119 Hamburg
http://www.diplomica-verlag.de, Hamburg 2017
Printed in Germany

Table of Contents

Chapter 1 **Introduction**	1
1.1 FSP	1
1.2 Microstructural zones in FSP	3
1.3 Capabilities of FSP	5
1.4 Tool Profiles	7
1.5 Applications of FSP	12
1.6 Advantages of FSP	13
1.7 Limitations of FSP	13
1.8 Need of the Present Study	14
1.9 Chapter Scheme	14
Chapter 2 **Literature Review**	16
2.1 Review of related literature	16
2.2 Research Gaps	31
2.3 Methodology	31
Chapter 3 **Experimentation**	33
3.1 Process Parameters and Procedure	33
3.2 Material	34
3.3 Equipments	35
3.4 FSP Tools	38
3.5 FSP Procedure	41
3.6 Characterization and Testing of Samples	42
Chapter 4 **Results and Discussion**	49
4.1 Microstructure Results	49
4.2 Results of Micro Hardness	53
4.3 Impact Strength Results	55
4.4 Rockwell Hardness Test Results	56
Chapter 5 **Conclusion and Future Scope**	59
5.1 Conclusions	59
5.2 Scope for Future Work	59
References	60

List of Tables

Table 3.1	Process Parameters	33
Table 3.2	Specimen Dimensions	33
Table 3.3	Chemical Composition of AL 6063	34
Table 3.4	Specification of CNC Vertical Milling Machine	36
Table 4.1	Microstructure Results	52
Table 4.2	Micro hardness Comparison Results	54
Table 4.3	Impact Strength Results	55
Table 4.4	Rockwell Hardness Results	56

List of Figures

Fig. 1.1	FSP Principle	1
Fig. 1.2	FSP Tool	2
Fig. 1.3	FSP Processed Specimen	2
Fig. 1.4	FSP Zones	4
Fig. 1.5	SEM micrographs showing the SiO2 particle dispersion in the SiO2/AZ61 composite prepared by FSP	5
Fig. 1.6	Surface Modification after processing	6
Fig. 1.7	FSP Tool Diagram of FSP tool profile design for show all views	8
Fig. 1.8	different tool pin profiles	9
Fig. 1.9	Plates processed with different tool pin profiles (a)-(d)	10
Fig. 1.10	Drawing of the FSP tool threaded pin	11
Fig. 3.1	Base Plate AL 6063	34
Fig. 3.2	CNC Vertical Milling Machine	35
Fig. 3.3	Fixtures Used in experimentations	37
Fig. 3.4	Fixture hold Al plate on bed of CNC vertical milling machine	38
Fig. 3.5	FSP Tool made from HSS	39
Fig. 3.6	Square tool used for FSP	40
Fig. 3.7	Drawing of FSP Tool	40
Fig. 3.8	Pentagonal tool pin profile	41
Fig. 3.9	FSP	41
Fig. 3.10	Optical Microscope	43
Fig. 3.11	AL sample mounted on plastic foil for microstructure examination	43
Fig. 3.12	Buffing Machine	44
Fig. 3.13	micro hardness measurement machine and mechanism	45
Fig. 3.14	Izod Impact Testing Machine	46
Fig. 3.15	AL 6063 sample prepared for impact testing	46
Fig. 3.16	Striking position and specimen fracture after izod test	47
Fig. 3.17	Pyramid shape indenter and indenter marks in processed zone	48
Fig. 3.18	Rockwell Hardness Testing Machine	48
Fig. 4.1	Microstructure of AL base plate	49
Fig. 4.2	Processed zone microstructure for pentagonal tool pin	50
Fig. 4.3	Microscopic image of microstructure of aluminium 6063 sample processed with square tool pin profile	50
Fig. 4.4	Microscopic image of microstructure of aluminium 6063 sample processed with threaded tool pin profile	51
Fig. 4.5	Microscopic image of microstructure of aluminium 6063 sample processed with circular tool pin profile	51
Fig. 4.6	Microstructure comparison with each tool pin profile processed specimen	53
Fig. 4.7	Micro hardness comparison of processed specimen with each tool pin profile	54
Fig. 4.8	Impact strength comparison of FSP processed sample with each tool pin profile	55
Fig. 4.9	comparison between Rockwell hardness of FSP processed specimen	57

CHAPTER 1
INTRODUCTION

1.1 FSP

Friction stir processing (FSP) is solid state process in which a non-consumable stirring (rotating) tool plunged into work piece up to half thickness, which causes intense plastic deformation, material mixing, and thermal exposure, resulting in refinement of micro structural, enhancement of mechanical properties and homogeneity of the processed (nugget) zone. The FSP technique has been successfully used for producing the fine-grained structure and surface composite, modifying the microstructure of materials, synthesizing the composite like metal-metal composites (MMC). The use of FSP generates significant frictional heating and intense plastic deformation, thereby resulting in the occurrence of dynamic recrystallization in the stirred zone (SZ). Although there is still a controversy about the grain-refinement mechanism in the SZ, it is generally believed that the grain refinement is due to dynamic recrystallization. Therefore, the factors influencing the nucleation and growth of the dynamic recrystallization will determine the resultant grain microstructure in the SZ. It has been demonstrated that the FSP parameters, tool geometry, material chemistry, workpiece temperature, vertical pressure, and active cooling exert a significant effect on the size of the recrystallized grains in the SZ. [In friction stir processing, a rotating tool with pin and shoulder is (made from material like HSS, mild steel etc.) inserted up to 1/2 thickness of work piece, and shoulder touches the workpiece surface as shown in following diagram]

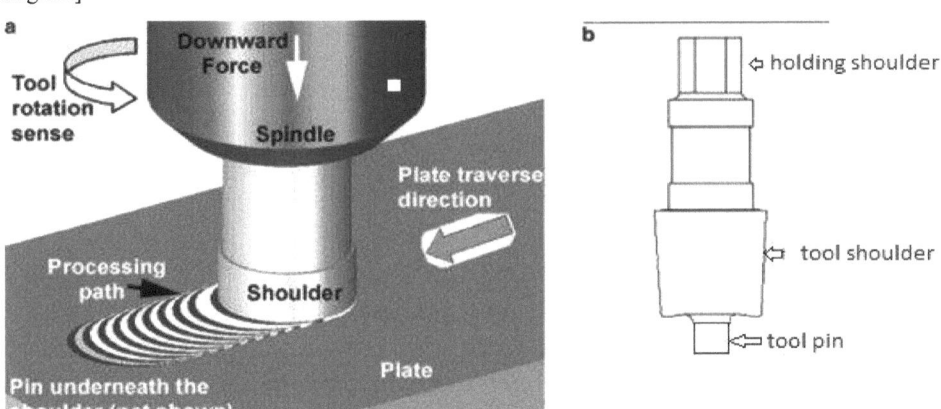

Fig. 1.1: FSP Principle (Mishra R.S *et al.,* 2005)

Following diagrams (Fig. 1.2) show the FSP tool and (Fig. 1.3) procesed Al plate sample which clears the process fundamentals.

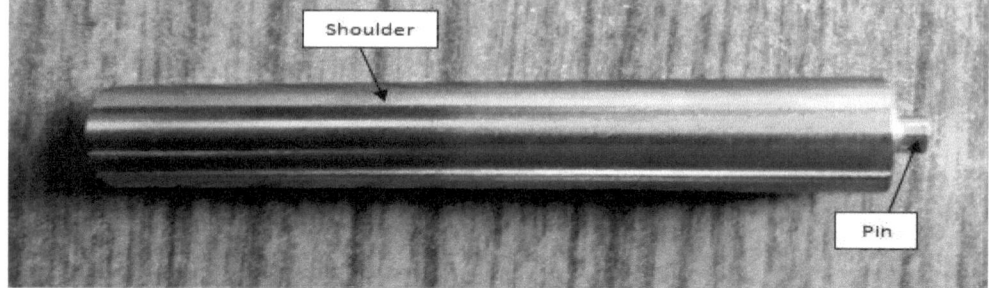

Fig. 1.2: FSP tool

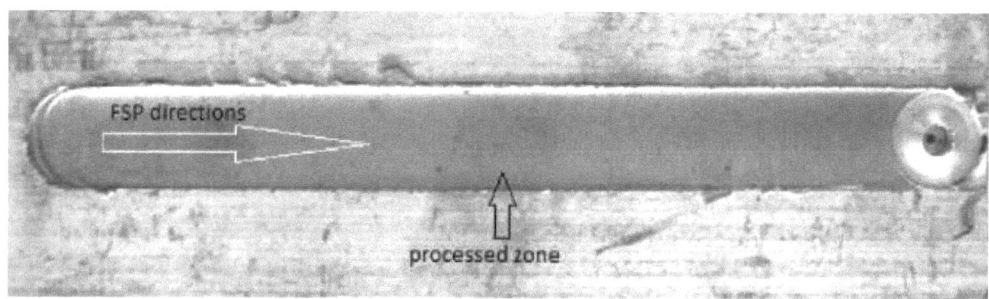

TMAZ=Thermo-mechanical zone
HAZ=Heat effected zone

Fig. 1.3: FSP processed specimen

The FSP technique is emerging as a very effective solid-state processing (material remains in plastic state) technique that can provide localized modification and control of microstructures of soft materials in the near-surface layers of processed metallic components. In the relatively short duration after its invention, increasing applications are being found for FSP in the fabrication, processing, making composites etc. Friction stir process is quite simple process which can well controlled by using numerical controlled machines. The use of FSP generates significant frictional heating and intense plastic deformation, thereby resulting in the occurrence of dynamic recrystallization in the stirred zone (SZ) or nugget zone (NZ).

Furthermore, the FSP technique has been used for the fabrication of a surface composite on aluminium substrate and the homogenization of powder metallurgy (PM) aluminium alloys, metal matrix composites, and cast aluminium alloys. Compared to other metalworking techniques, in FSP, the major processing parameters are the tool rotating rate, the tool traversing speed and proper tool pin profile there are different ^{types} of tool profiles. The intense plastic deformation around the tool and the friction between the tool and the work piece both contribute to the temperature increase in the stirred zone (SZ). SZ is processing zone where tool starting stirring (rotating). Generally, an increase in the ratio of the rotation rate to the traversing speed will increase the peak temperature in FSP. In the present thesis it is try to find out proper tool pin profile in friction stir processing. In the present study, the effect of the tool pin profile on microstructure and mechanical properties is to be examined that how these profile effect microstructure and mechanical properties of aluminium 6063.

1.2 Microstructural zones in FSP

During processing material divided into following type of zones which arises due to mechanical action and frictional heat of tool on material microstructure. The microstructure can be broken up into the following zones (Fratini and Buffa, 2005)

a. Parent material

It is unaffected zone which does not goes under any deformation and remains same before and after processing This is the zone of parent material away from processed metal which is not affected by the heat flux in terms of microstructure or mechanical properties. In this zone no material deformation occurs.

b. Heat affected zone (HAZ)

In this region the material undergoes a thermal cycle which leads to modified microstructure and/or mechanical properties. This zone retains the same grain structure as the parent materials. However, no plastic deformation occurs in this area but it is affected due to heat dissipated in nugget zone

c. Thermo-mechanically affected zone (TMAZ)

In this zone, the material undergoes plastic deformation by the tool, and the heat flux also exerts some influence on the material. TMAZ is produced by friction between the tool shoulder and the top surface of plate, as well as plastic deformation of the material in contact with the tool. In case of aluminium, no recrystallization is observed in this zone; however extensive deformation is present in this region

d. Nugget or processed zone (NZ)

The recrystallized area below the tool shoulder is given a separate category, as it has different grain structure. In this zone, the original grain and sub grain boundaries appear to be replaced with fine, equi-axed recrystallized grains characterized by a nominal dimension of few micro meters. In this region intense plastic deformation and frictional heating during FSP result in recrystallized fine-grained microstructure.

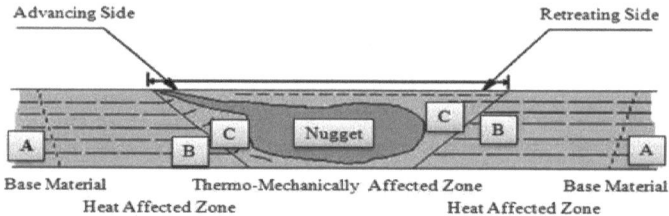

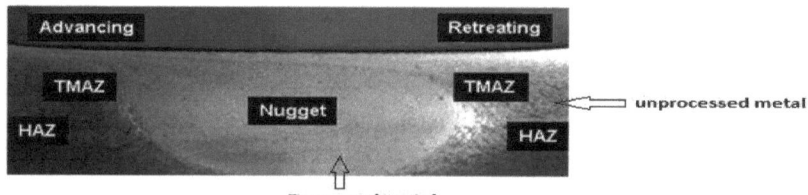

Fig. 1.4: FSP zones

1.3 Capabilities of FSP

1.3.1 Fabrication of composite by friction-stir processing

Friction stir process show better result during fabrication of composites, the use of the FSP technique results in the intense plastic deformation and mixing of material in processed zone; results in fabrication of composites, it is possible to incorporate the ceramic particles into the metallic substrate plate, to form the surface composites. (Mishra *et al.,*) reported the first result on the fabrication of a SiCp-Al surface composite via FSP. The SiC powder was added into a small amount of methanol and mixed, and was then applied to the surface of the plates, to form a uniform thin SiC particle layer. Due to stirring action of tool silicon powder well mixed and fabricated in base metal. In general cases aluminium plate used as base plate when silicon powder is matrix phase.

The aluminium plates with a preplaced SiC particle layer were subjected to FSP. With the optimized tool and pin profile design and processing parameters, a composite layer of ~100 lm, with well distributed particles and a good bonding with the aluminium substrate, was generated on the substrates of 5083Al and A356 aluminium alloys. By adjusting the FSP parameters, 5 to 27 vol. of the SiC particles could be incorporated into the aluminium matrix. Optimum tool design and process parameter are responsible for properties of composites to be fabricated Slower feed rates means more stirring action per unit area, more stirring means better material flow rate. So to faster speed does not show best result as compare to slower ones.

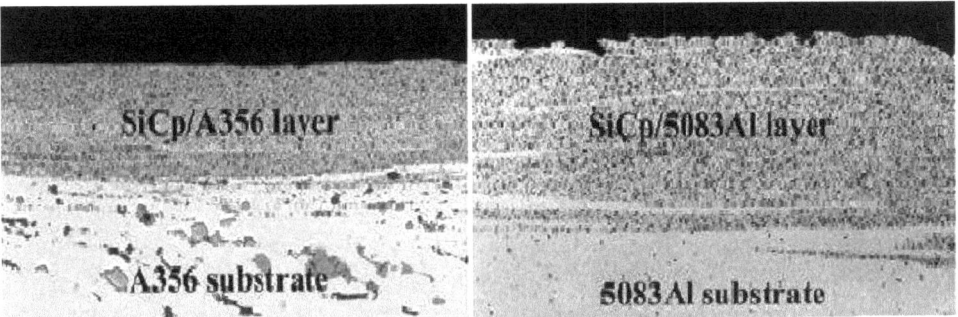

Fig. 1.5: SEM micrographs showing the SiO2 particle dispersion in the SiO2/AZ61 composite prepared by FSP.

Above diagram shows the capabilities of FSP in fabrication of composites. Right hand diagram shows that matrix particle homogeneously spreaded all over the volume of base metal i.e. aluminium. Also porosity also reduces and metal composite become very fine. Friction stir process is suitable for aluminium composites fabrications. (Mishra R.S *et al.*, 2005)

1.3.2 Friction-stir microstructural modification

FSP is unique process for surface modification. Now question is that why? Answer is that because stirring action of tool flow the metal and fill various casting voids and refines grain boundaries and grain size due to recrystallization in nugget zone. During FSP, the rotating pin with a threaded design produces an intense breaking and mixing effect in the processed zone, thereby creating a fine, uniform, and dense structure. Therefore, FSP can be developed as a generic tool for modifying the microstructure of heterogeneous metallic materials such as cast alloys, metal matrix composites, and nano phase aluminium alloys prepared through the PM technique. Alloys of Al-Si-Mg (Cu) are widely used to cast high strength components in the aerospace and automobile industries, because they have good cast ability and can be strengthened by artificial aging. The as-cast structure of Al-Si-Mg (Cu) alloys is characterized by porosity, coarse Si particles, and coarse primary aluminium dendrites. These microstructure features limit the mechanical properties of cast alloys, in particular, toughness and fatigue resistance. Eutectic modifiers and high-temperature heat treatment are widely used to refine the microstructure of cast Al-Si alloys, to enhance the mechanical properties of the castings. However, approaches can heal the casting porosity effectively and redistribute the Si particles uniformly into the aluminium matrix. (Mishra R.S *et al.*, 2005)

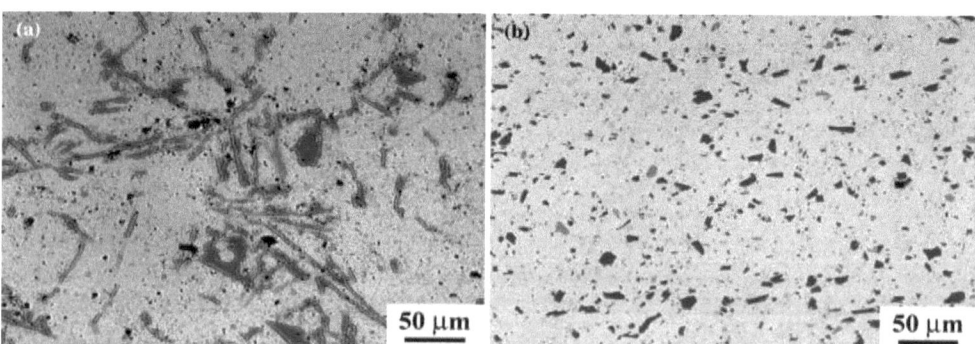

Fig. 1.6: Surface modification after processing (silicon-Al composite)

Above diagram shows the capabilities of FSP in modification in surface of metal. FSP tool can remove casting defects and porosity hence surface of metal become fine and strength will also increases as shown in above diagrams surface become more porous and defects free. Reason is that stirring action of tool melts and flow the material of metal hence small casting defects filled due to flow of material and material microstructure become fine grained and homogeneous.

Interest in friction stir processing rapidly growing as it is refining materials microstructure and internal properties through single pass of non-consumable tool without any addition of alloying elements for aluminium and its alloys. Aluminium is the most popular metal that is widely used. About 85% of aluminium is used for wrought products, for example rolled plate, foils and extrusions. Aluminium has light weight, resistance to corrosion and has low melting point but its severe limitation is the difficulty associated with welding of aluminium/aluminium alloy structures many times aluminium failed to sustain heavier loads and also having poor surface hardness.

The research work carried out by some of the investigators reveals that friction stir processing technique can be successfully used for micro structural modification, enhancement of surface properties such as micro hardness, wear resistance etc. of aluminium or its alloys. Materials processed by friction stir process used in complex aerospace part where high specific strength per unit area required, best suited for manufacturing composites. In the light of above mentioned facts mostly circular tool profile is used which show better results on aluminium composites. Now, the effort shall be made to find out the effect of other tool profiles either than circular tool profile. If other tool profile show better result than circular then these tool profile can be used instead of circular tool profile So effort is made to find out effect of four tool profiles on aluminium 6063 so that desired tool can be selected for desired results.

1.4 Tool profiles
The tool geometry plays a critical role in material flow and in turn governs the traverse rate at which FSP can be conducted. Tool has two primary functions
1. Localized heating
2. Material flow

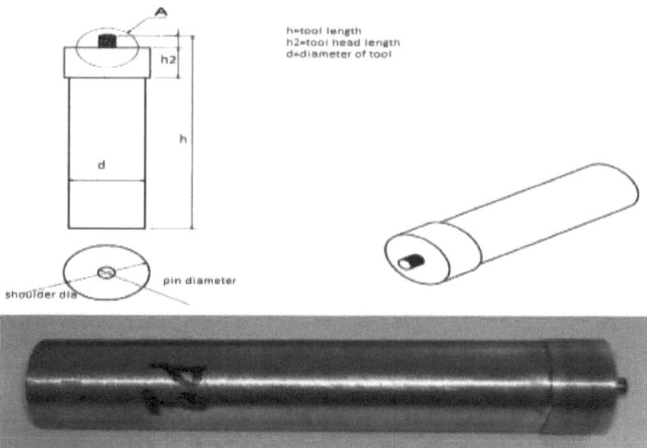

Fig. 1.7: FSP tool diagram of FSP tool profile design for show all views

In first function tool provide heat which melts the material and this heat is due to centrifugal force friction between tool shoulder and work piece. The tool is plunged till the shoulder touches the workpiece. The friction between the shoulder and workpiece results in the biggest component of heating. From the heating aspect, the relative size of pin and shoulder is important, and the other design features are not critical. Now it is clear that tool pin and shoulder are responsible for localized heating the shoulder also provides confinement for the heated volume of material. The second function of the tool is to _stir' and _move' the material. The uniformity of refined fine grained microstructure and mechanical properties as well as process loads. In present research it is try to find out effect of tools profiles (Circular tool pin profile, Pentagonal profile, Threaded tool profile, Square tool profile)

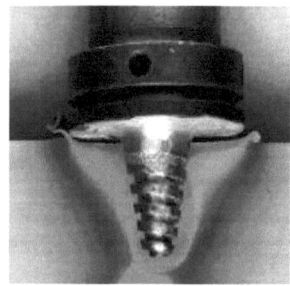

threaded pin profile

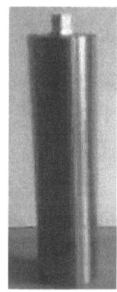

square pin profile

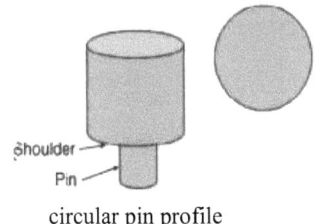 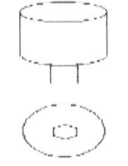

circular pin profile   pentagonal pin profile

Fig. 1.8: different tool pin profiles

Following are specimen processed with each profile

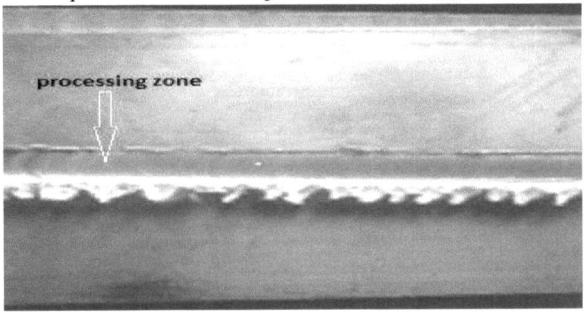

(a) square

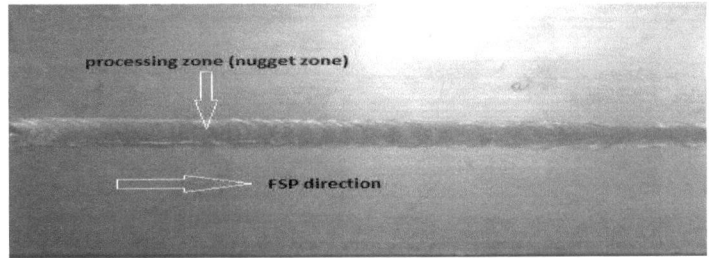

(b) pentagonal

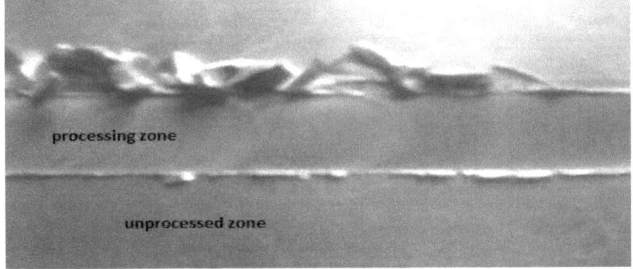

(c) circular

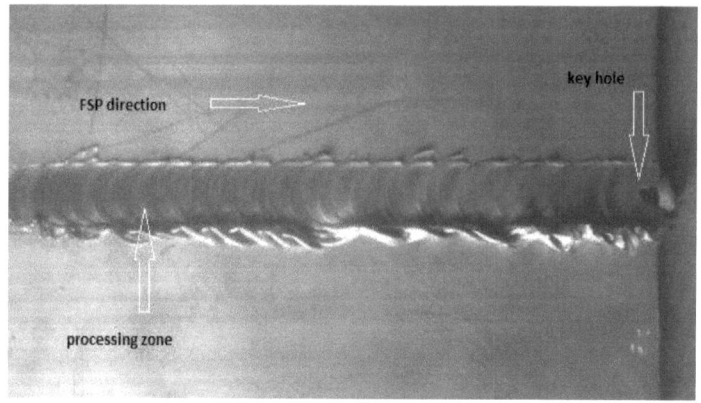

(d) threaded

Fig. 1.9 Plates processed with different tool pin profiles

All above diagrams shows four plates processed with four tool pin profiles. It is show that every plate having different effects with each profile i.e. type of chips etc.

1.4.1 Circular tool pin profile

Mostly used tool profile is circular tool pin profile as shown in (Fig. 1.6). It simpler in design and can be machined easily on center lathe. Tool pin length varied according to the thickness of work piece, it should be ½ thickness of work piece and shoulder should touch with surface of material to be processed as shown in first diagram Now using the working material get deformed and melts during single pass of tool and after pass complete material solidified during process with circular tool profile because to pin is round less power is required and also less wear tear of pin than that of pentagonal tool profile.

1.4.2 Square tool pin profile

With increasing experience and some improvement in understanding of material flow, the tool geometry has evolved significantly. Complex features have been added to alter material flow, mixing and reduce process load. Another tool pin profile is square tool pin profile which having pentagonal base of pin tip, remaining dimensions same as circular profile so we can say that square tool profile is modified form of circular tool profile. Four sharp edges of square tool profile are responsible for material flow rate. If four edges of tool profile replace with six edges a new profile will formed known as pentagonal tool pin profile. These edges are more responsible in tool performance due material flow rate and heat generation during stirring of tool in nugget zone

(NZ). But these tool profiles have a disadvantage that more wear rate of tool profile edges than circular tool pin profile.

1.4.3 Threaded tool profile

Another tool pin profile is threaded tool pin profile that pins for both tools are shaped as a threading that displaces less material than a cylindrical tool of the same root diameter. Typically, reduces displaced volume by about 60%, machining load also minimum than other profiles.

Fig. 1.10: Drawing of the FSP tool threaded pin (Mishra R.S *et al* 2005)

1.4.4 Pentagonal tool profile

Pentagonal tool profile same as square tool profile except that there are 6 sharp edges on pin. These sharp edges are responsible for more stirring action in nugget zone. Main difference is that circumference volume is more and sharp edges are more responsible for material flow rate but wear tear is more than other profiles. Tool geometry is the most influential aspect of process development. The tool geometry plays a critical role in material flow and in turn governs the traverse rate at which FSW can be conducted. An FSP tool consists of a shoulder and a pin. As mentioned earlier, the tool has two primary functions: (a) localized heating, and (b) material flow. In the initial stage of tool plunge, the heating results primarily from the friction between pin and work piece. The tool is plunged till the shoulder touches the work piece. The friction between the shoulder and work piece results in the biggest component of heating. From the heating aspect, the relative size of pin and shoulder is important, and the other design features are not critical. The shoulder also provides confinement for the heated volume of material. The second function of the tool is to _stir and _move the material. The uniformity of microstructure and properties as well as process loads are governed by the tool design. Generally a concave shoulder and threaded cylindrical pins are used. With increasing experience and some for multi pass FSP, conventional cylindrical threaded pin resulted in excessive thinning of the top sheet, leading to significantly reduced bend properties (Mishra R.S *et al.*, 2005).

1.5 Applications of FSP

Friction stir processing is used in number of industries as described below

☐ Casting: Metallic parts produced by casting are comparatively inexpensive, but are often subject to metallurgical flaws like porosity and microstructural defects. Friction stir processing can be used to introduce a wrought microstructure into a cast component and eliminate many of the defects. By vigorously stirring a cast metal part to homogenize it and reduce the grain size, the ductility and strength are increased.

☐ Powder metallurgy: Friction stir processing can also be used to improve the microstructural properties of powder metal objects. In particular, when dealing with aluminium powder metal alloys, the aluminium oxide film on the surface of each granule is detrimental to the ductility, fatigue properties and fracture toughness of the work piece. While conventional techniques for removing this film include forging and extrusion, friction stir processing is suited for situations where localized treatment is desired.

☐ Shipbuilding and Marine Industries: The shipbuilding and marine industries are two of the first industrial sectors, which have adopted FSP process for commercial applications (Delany *et al.*, 2007)

☐ Aerospace Industry: At present the aerospace industry is using FSW/FSP for welding of prototype parts. Opportunities exist to weld skins to spars, ribs and stringers for use in military and civilian aircraft. This offers significant advantages compared to riveting and machining from solid, such as reduced manufacturing costs and weight savings. Longitudinal and circumferential lap joints of Al alloy fuel tanks for space vehicles have been friction stir welded and successfully launched in commercial flights. The process could also be used to increase the size of commercially available sheets by welding them before forming. FSP processed part can also show more hardness which can be used instead of unprocessed aluminium. FSP processed aluminium can be used instead of base (unprocessed) aluminium.

1.6 Advantages of FSP

The advantages of the FSP technique compared to these other processing techniques are summarized.

- FSP is a relatively simple processing technique with a one step processing that produces a fine-grained microstructure. Other processing techniques are relatively complex and time-consuming and lead to increased material cost. For example, TMT involves solution treatment, averaging, multiple-pass warm rolling with intermittent reheating, and a recrystallization treatment. For ECAP, at least 4 to 6 passes are required to achieve microstructural refinement.

- FSP does not reduce the thickness of the processed plates; therefore, it is possible to achieve the super plastic forming of thick plates. Depending on the length of the tool pin, the fine-grained microstructure can be achieved in the plates with a thickness of up to tens of milli meters via FSP. In uniform elongations (El.) up to 500 pct in a 12-mm-thick 7475Al unpublished research, Mahoney *et al.* have illustrated plate prepared via FSP.

- A local processing can be achieved via FSP. In this case, it is possible to produce a local fine-grained microstructure in a region that will undergo super plastic deformation. A concept of selective super plastic forming has been suggested and demonstrated by means of FSP. comparison, other techniques cannot produce microstructural refinement on a selective basis.

- The fine and equi axed grains with high ratio of high-angle boundaries in FSP aluminium alloys resulted in the generation of high-strain-rate super plasticity and low temperature super plasticity.

1.7 Limitations of FSP

Although there are many advantages but there are some limitations associated with friction stir process as outlined below (Skinner and Edward 2003).

(i) Processing speeds are somewhat slower.
(ii) Work pieces need to be rigidly clamped for proper processing.
(iii) Support is required on back side of work pieces.
(iv) There is a keyhole at the end of each processed specimen..
(v) It is difficult to understand the interactions between tool and workpiece material.
(vi) Difficult to develop tool material to extend the technology to materials with high melting point.

(vii)   It is complex to model material flow and thermo-mechanical interaction.

1.8 Need of the present study

Interest in friction stir processing rapidly growing as it is refining materials microstructure and internal properties through single pass of non-consumable tool without any addition of alloying elements for aluminium and its alloys. Aluminium is the most popular metal that is widely used. About 85% of aluminium is used for wrought products, for example rolled plate, foils and extrusions. Aluminium has light weight, resistance to corrosion and has low melting point but its severe limitation is the difficulty associated with welding of aluminium/aluminium alloy structures many times aluminium failed to sustain heavier loads and also having poor surface hardness. The research work carried out by some of the investigators reveals that friction stir processing technique can be successfully used for micro structural modification; enhancement of surface properties such as micro hardness, wear resistance etc. of aluminium or its alloys by using mostly circular tool pin profile. In present research four tool pin profile were used to find out effect of each pin profile on microstructure and mechanical properties of Al 6063 for objective that can other tool pin profile also improve mechanical properties.

1.9 Chapter scheme

Chapter 1 Introduction

It introduce the basic working principle of friction stir process and some useful capabilities of FSP according to research papers and review paper of FSP.It also includes the various tool profiles and material (Al6063) which used in this research.

Chapter 2 Literature review

It includes the past research on FSP.It help to find out previous research gaps which are very useful to continue the research. It also shows various research gaps based on literature survey.

Chapter 3 Experimentation

It defines and explains the research and then help to clear objectives. It also shows overall experimentation procedure, with all the process parameter used during research.

Chapter 4 Result and discussion

It show overall output of experimentations which may be positive or may be negative in form of results. Results are also discussed in this chapter.

Chapter 5 Conclusion and future scope

It show final result discussed and gives conclusion of result and experimental study with scope in future.

CHAPTER 2
LITRATURE REVIEW

2.1 Review of related literature

Following is literature review of friction sir processing based on research paper and journal of friction stir process.

- **A. Shafiei (2008)** studied, a new processing technique, friction stir processing (FSP) was attempted to incorporate nano-sized Al2O3 into 6082 aluminium alloy to form particulate composite surface layer. Samples were subjected to various numbers of FSP passes from one to four, with and without Al2O3 powder. Microstructural observations were carried out by employing optical and scanning electron microscopy (SEM) of the cross sections both parallel and perpendicular to the tool traverse direction. Mechanical properties include micro hardness and wear resistance, were evaluated in detail. The results show that the increasing in number of FSP passes causes a more uniform in distribution of nano-sized alumina particles. The micro hardness of the surface improves by three times as compared to that of the as-received Al alloy. A significant improvement in wear resistance in the nano-composite surfaced Al In the present investigation, the Al/Al2O3 surface nano composites were successfully fabricated by the FSP. The microstructure, micro hardness and tribological behaviour were evaluated by a view of nano-composite layer produced by four FSP passes. The FSP with the nano-sized Al2O3 particles more effectively reduce the grain size of the 6082 Al matrix in which some grains are less than 300 nm for the surface nano-composite layer produced by four FSP passes. It is considered that the pinning effect by the nano-sized Al2O3 particles retarded the grain growth of the dynamically recrystallized grains of Al matrix. The micro hardness of the Al/Al2O3 surface nano-composites increases significantly with increasing number of FSP passes. The maximum micro hardness for the surface composites is 295Hv, while that of the samples treated by the FSP without Al2O3 particles and the as-received Al are 67 and 110Hv, respectively. The high micro hardness of Al/Al2O3 surface nano-composite can be attributed to the presence of nano-sized Al2O3 particles, which contribute significantly to the strength through the Orowan mechanism, as well as the ultrafine grain size of the Al matrix.

- **Bhalchandra *et al.*, (2012)** show that Friction Stir Processing (FSP) is a promising technique to develop surface composite. The aim of the present study is to develop defect free surface composite of Al 5083 alloy reinforced with TiC particles and investigate the particle distribution in the matrix, mechanical properties and wear behaviour of the composites. Microstructural observations were carried out by using optical and scanning electron microscopy (SEM). The microstructural studies revealed that distribution of particles were more uniform in samples subjected to double pass than the single pass FSP. The micro hardness profiles along top surface and across the cross section of the processed samples were evaluated. The average hardness along the top surface was found to increase by 27.27%, as compared to that of the base metal (88Hv). The particles were incorporated maximum average depth about 250μm in the surface composite. The slurry erosion tests revealed that the wear rate was highly reduced in case of double pass FSP samples as compared to base metal and single pass FSP composite Surface composite developed by single pass FSP with groove design, the average hardness along the top surface was found to increase by 22.72% as compared to that of the base metal whereas the in case of surface composite developed by double pass FSP in same and opposite direction, the average hardness along the top surface was found to increase by 25% and 27.27% respectively, as compared to that of the base metal. The maximum average depth of surface composite was found to be 250μm in hole-design.

- **Buff *et al.*, (2008)** studied that friction stir welding and processing both a thermal flux and a mechanical action are exerted on the material determining metallurgical evolutions, changes in the mechanical behaviour and a complex residual stress state. In the paper, the metallurgical changes are examined through numerical simulation and experiments to highlight and distinguish the effects of thermal and mechanical loadings. A particular focus is made on the residual stresses generated during the stir processing of AA7075-T6 aluminium blanks. The predictions of FE model are validated by experimental measurements. Lastly, this paper presents an in-process quenching of the processed blanks for improved mechanical properties and microstructure.

- **B. Zahmatkesh *et al.*, (2010)** studied that FSP is a new approach was used to produce Al-10%Al2O3 surface nano composite on Al2024 substrate. This novel approach involved air plasma spraying of Al-10%Al2O3 powder to produce Al- 10%Al2O3 coating on substrate. The coated material was then subjected to friction stir processing (FSP) to distribute Al2O3 particles into the substrate. Microstructure and mechanical properties of samples were investigated by optical microscopy (OM), scanning electron microscopy (SEM), micro hardness and wear measurements. As a result, it was found that the Al2O3 particles were distributed uniformly inside the substrate with an average penetration depth of about 600 μm. The surface nano composites produced in this way had excellent bonding with the substrate. The micro-hardness of the surface nano composite was ~230 Hv, much higher than ~90 Hv for Al2024 substrate. The surface nano composites also exhibited lower friction coefficient and wear rate. It was found that the addition of Al2O3 nanoparticles to the Al2024 matrix alloy affect the mechanism of wear.

- **B.C. Liechty *et al.*, (2008)** examine flow behaviour in friction stir processing of metals is investigated by tracking the motion of discrete particles and grid deformation in a plasticine workpiece. Mechanical and thermal similarities between plasticine and metals are discussed. These similarities along with ease of particle/grid setup and post-processing examination make plasticine an ideal analogue material for friction stir processing. Uniformly spaced steel particles are arranged along lines parallel to the tool feed direction at the mid-pin depth in the plasticine. During friction stir processing of the plasticine, the forward motion of the tool is stopped nearly instantaneously in order to freeze the flow of material around the tool. X-ray images of particle streamlines are acquired after processing. The particle spacing before and after processing is used to calculate velocities, strains, and strain-rates along streamlines. Results from the particle streamlines show substantial slip at the material/pin interface. The maximum velocity of the material in contact with the pin is only 6.0–7.0% of the pin speed. Generally, material initially at the far advancing side of the pin is compressed, while the remaining material near the pin is stretched during processing. The highest strain and strain-rate are 4.4 and 1.3 s^{-1}, respectively, which occurs in the material that contacts the pin. The gross flow of material around the tool pin is also observed from contrasting colors of the plasticine, which are arranged as markers in several layers parallel to the feed direction. Markers at the mid-pin depth in the processed zone exhibit significant deformation and irregular flow.

- **Chang *et al.*, (2006)** studied and show that Mg-AZ31 based composites with 10{20%vol. nano-sized ZrO2 and 5{10%vol. nano-sized SiO2 particles were fabricated by friction stir processing (FSP). The clusters of the nano-ZrO2 and nano-SiO2 particles, measuring 180–300nm in average, were relatively uniformly dispersed. The average grain size of the Mg matrixes of the composites varied within 2–4 mm after four FSP passes. No evident interfacial particles, measuring 180–300nm in average, were relatively uniformly dispersed. The average grain size of the Mg matrixes of the composites varied within 2–4 mm after four FSP passes. No evident interfacial product between the ZrO2 particles and Mg matrix was found during the FSP mixing

ZrO2 into Mg-AZ31. However, significant chemical reactions occurred at the Mg/SiO2 interface to form the Mg2Si phase. Friction stir processing successfully fabricated bulk Mg-AZ31 based composites with 10{20%vol. of nano- ZrO2 particles and 5{10%vol. of nano-SiO2 particles. The distribution of the 20 nm nano-particles after four FSP passes resulted in satisfactorily uniform distribution.

- **Chen *et al.*, (2009)** show that Magnesium alloys present a great potential as structural materials in the aerospace and automobile industries because of several advantages, such as low density, high specific strength and good machinability. The microstructural evolution characteristics of thermo-mechanically affected zone were investigated during friction stir processing (FSP) of the thixo formed AZ91D alloy. Simultaneously, an Al-rich surface layer was prepared by combination of Al powder using FSP method. The results indicate that the dynamic recrystallization and mechanical separation (including splitting and fracture of the primary grains) are the main mechanisms of grain refinement. For the thixo formed alloy, the operation efficiency of these mechanisms is less than that of the permanent mould casting AZ91D alloy, thus its microstructural evolution is relatively slow and the resulting grain size is relatively large. These are attributed to the differences in their original microstructures. The Al-rich surface layer can obviously improve the corrosion resistance in NaCl aqueous solution. A proper solution heat treatment can further increase the corrosion resistance. In order to improve corrosion resistance, increasing the amount and improving the distribution uniformity of the Al-rich phase are more effective than increasing the Al solubility in the matrix. Compared with the PMC alloy, the microstructural evolution of the TF alloy is slower because the operation efficiency of the grain refinement mechanisms is relatively low. The resulting grain size of the TF alloy is also larger at a given stir pass. But this difference decreases with the increase of stir pass.

- **Cavaliere *et al.*, (2005)** studied that the mechanical and microstructural properties of 7075 aluminium alloy resulting from Friction Stir Processing (FSP), into sheets of 7 mm thickness, were analysed in the present study. The sheets were processed perpendicularly to the rolling direction; the tensile mechanical properties were evaluated at room temperature in the transverse and longitudinal directions with respect to the processing one. Tensile tests were also performed at higher temperatures and different strain rates in the nugget zone, in order to analyse the superplastic properties of the recrystallized material and to observe the differences from the parent material as a function of the strong grain refinement due to the Friction Stir Process. The high temperature behaviour of the material was studied, in the parallel direction, by means of tensile tests in the temperature and strain rate ranges of 150–500 8C respectively, electron microscopy (FEGSEM) observations were carried out to investigate more closely the fracture surfaces of the specimens tested at different temperatures and strain rates.

- **Douglas *et al.*, (2007)** in this research, equations are developed to model the rate of heat input from different geometries of friction stir processing (FSP) tools. The model is then compared with actual heat input obtained from embedded thermocouples within the stirred region. The cooling curves obtained from the thermocouple data are then applied to the Derby–Ashby model for high angle grain boundary migration to predict the final grain size of a bulk sample produced by the friction stirring method. Submerged friction stir processing (SFSP) is introduced as a way of increasing the cooling rate of the bulk samples in an attempt to decrease the grain size. Microstructures obtained from both FSP and SFSP are characterized using transmission electron microscopy.

- **Darras *et al.*, (2007)** studied recently friction stir processing (FSP) has emerged as an effective tool for enhancing sheet metal properties through microstructure modification. Significant grain refinement and homogenization can be achieved in a single FSP pass leading to improved formability, especially at elevated temperatures. FSP is a solid-state process where the material within the processed zone undergoes intense plastic deformation resulting in dynamically recrystallized grain structure. Most of the research conducted on FSP focuses on aluminium alloys. Despite the potential weight reduction that can be achieved using magnesium alloys, very little is reported on FSP of magnesium alloys. In this work, we examine the possibility of using FSP to modify the microstructure and properties of commercial AZ31B-H24 magnesium alloy sheets. The effect of various process parameters on thermal histories, resulting microstructure and properties are investigated. Preliminary results are promising and it is shown that FSP leads to finer and more homogenized grain structure. The preliminary results on FSP of AZ31 magnesium alloy are promising; grain refinement and homogenization of the microstructure are achieved in a single FSP pass. The thermal histories presented in these studies give useful results on the peak temperature, cooling and heating rates which are critical to control and optimize the process.

- **Ehab *et al.*, (2009)** studied that Commercial 5083 Al rolled plates were subjected to friction stir processing (FSP) with a tool rotational speed of 430 rpm and a traverse feed rate of 90 mm/min. This treatment resulted in a fine grained microstructure of 1.6 lm and an average misorientation angle of 24. Ductility was measured using tensile elongations at a temperature of 250 C at three strain rates, and demonstrated that a decrease in grain size resulted in significantly enhanced ductility and lower forming loads. The ductility of the friction stir processed material was enhanced by a factor ranging from 2.6 to 5 compared to the ductility of the as received material, in the range of the strain rates tested. The strain rate sensitivity of the processed material is 0.33 while for the as received, it is 0.018. The deformation mechanism, in the fine-grained specimens is mainly controlled by solute drag creep, though the contribution of grain boundary sliding to the deformation process cannot be overlooked. Both mechanisms led to significant flow localization and simultaneous cavity formation Friction stir processing is capable of producing dynamically recrystallized fine microstructure. The grain morphology shifted from elongated grains in the base metal to fine equi axed morphology (1.6 lm) with high angle grain boundaries in the nugget zone.

- **Essam *et al.,* (2010)** studied that Aluminium-base hybrid composites reinforced with mixtures of SiC and Al2O3 particles 1.25_minaverage size have been fabricated on an A 1050-H24 aluminium plate by friction stir processing (FSP) and their wear resistance has been investigated as a function of relative weight ratios of the particles. A mixture of SiC and Al2O3 powders of different weight ratios was packed into a groove of 3mmwidth and 1.5mm depth cut on the aluminium plate, and covered with an aluminium sheet 2mm thick. A FSP tool of square probe shape, rotated at a speed of 1500 rpm, was plunged into the plate through the cover sheet and the groove, and moved along the groove at a travelling speed of 1.66 mm/s. After the hybrid composite was fabricated on the Al plate, the homogeneity of the particles distribution inside the Al matrix has been evaluated from the macro/microstructure and hardness distribution. Moreover, the wear characteristics of the resulted hybrid composites were evaluated using a ball-on-disc wear tester at room temperatures at normal loads of 2, 5, and 10 N. As a result, it was found that the reinforcement particles were distributed homogenously inside the nugget zone without any defects except some voids that appeared around the Al2O3 particles. The average hardness decreased with increasing the relative content of Al2O3 particles. Regarding the wear characteristics, the wear volume losses of the hybrid composites depended on the applied load and the relative ratio of SiC and Al2O3 particles. The hybrid composite of 80% SiC + 20% Al2O3 showed superior wear resistance to 100% SiC and Al2O3 or any other hybrid ratios at a normal load of 5N, while the wear resistance was insensitive to the reinforcement ceramic type and was very close to the unreinforced FSP sample at a normal load of 10 N.

- **Hsu et al., (2006)** objective for this study is to produce fully dense intermetallic-reinforced aluminium composites by the use of the FSP technique. Friction stir processing (FSP) is applied to produce intermetallic-reinforced aluminium matrix composites from elemental powder mixtures of Al–Cu and Al–Ti. The intermetallic phases are identified as Al2Cu and Al3Ti, which are formed in situ during FSP. The volume fraction of the intermetallic phases in the in situ composites may reach as high as 0.5. The composites produced by FSP are fully dense with high strength, and the composite strength increases with the reinforcement content. This work has demonstrated that intermetallic-reinforced aluminium matrix composites with ultrafine-grained structure can be fabricated in situ by FSP. The intermetallic particles are distributed uniformly in the composites. The volume fraction of the intermetallic phases in the in situ composites may reach as high as 0.5. The composite thus produced are fully dense with improved strength, which increases with the reinforcement content.

- **Johannes et al., (2007)** search that AA5083 aluminium is a material of choice for superplastic automotive and aerospace aluminium panels. The cost of superplastic grade aluminium sheet, despite some reductions in recent years, remains relatively high. Continuous strip casting holds promise for low cost superplastic sheet. A continuously cast AA5083 aluminium sheet was friction stir processed and cold rolled in order to compare the properties with the as-cast and rolled material. It was found that the inclusion of the FSP step refined the recrystallized grain size, increased the elongations, and lowered the flow stresses.

- **Jerome et al., (2012)** studied Friction Stir Processing (FSP) is a promising technique to develop surface composite. The aim of the present study is to develop defect free surface composite of Al 5083 alloy reinforced with TiC particles and investigate the particle distribution in the matrix, mechanical properties and wear behaviour of the composites. Microstructural observations were carried out by using optical and scanning electron microscopy (SEM). The microstructural studies revealed that distribution of particles were more uniform in samples subjected to double pass than the single pass FSP. The micro hardness profiles along top surface and across the cross section of the processed samples were evaluated. The average hardness along the top surface was found to increase by 27.27%, as compared to that of the base metal (88Hv). The particles were incorporated maximum average depth about 250µm in the surface composite.

The slurry erosion tests revealed that the wear rate was highly reduced in case of double pass FSP samples as compared to base metal and single pass Friction Stir Processed composite. Surface composite developed by single pass FSP with groove design, the average hardness along the top surface was found to increase by 22.72% as compared to that of the base metal whereas the in case of surface composite developed by double pass FSP in same and opposite direction, the average hardness along the top surface was found to increase by 25% and 27.27% respectively, as compared to that of the base metal. The maximum average depth of surface composite was found to be 250μm in hole-design.

- **Kurt et al., (2010)** this study, SiC particles were incorporated by using Friction Stir Processing (FSP), into the commercially pure aluminium to form particulate surface layers. Samples were subjected to the various tool rotating and traverse rates with and without SiC powders. Microstructural observations were carried out by employing optical microscopy of the modified surfaces. Mechanical properties like hardness and plate bending were also evaluated. The results showed that increasing rotating and traverse rate caused a more uniform distribution of SiC particles. The hardness of produced composite surfaces was improved by three times as compared to that of base aluminium. Bending strength of the produced metal matrix composite was significantly higher than processed plain specimen and untreated base metal. Following results are obtained that FSP was an appropriate method to modify the microstructure and mechanical properties of 1050 Al-alloy. In general, FSP decreased the grain size and increased the hardness of processed material. Secondly increased rotation speed and low travelling speeds caused more heat input which affects the thickness of the surface layer, grain size and distribution of the precipitates and reinforcing particles. A good dispersion of SiCp can be obtained for the composite layer produced by ω= 1000rpm and v = 20 mm/min. Good interfacial conditions between particles and base metal can be formed during this solid-state process which avoids the chemical reactions on the interface.

- **Karthikeyan (2009)** studied and search that the surfaces of cast A319 alloy plates of nominal composition (wt.%): Al – 5.2 Si – 2.51 Cu were subjected to single stir Friction Stir Processing (FSP) with a view to decreasing the grain size and porosity level and improving the mechanical properties. Three traverses feed rates and five tool rotational speeds were employed. For certain combinations of the variables, the processed alloy displayed an increase in value of around 50% in tensile strength and 20% in micro hardness compared with those of the as-cast alloy. The ductility of the processed alloy had increased by a factor which ranged from 1.5 to 5. Optical and scanning electron microscopy revealed that FSP reduces the size of the second phase particles, which contributes to the improvements in mechanical properties. FSP increased the yield strength of the starting material by about 13% (maximum), while the enhancement in the tensile strength varied between 20% and 50%.In most of the processed specimens the hardness was greater, with the maximum increase being about 20%.

- **Liu *et al.*, (2010)** search in which he gives that commercial 2219Al–T6 alloy plates were friction stir processed at a rotation rate of 400rpm and a traverse speed of 100mm/min in water and air, producing two fine-grained 2219Al samples with average grain sizes of 1.0 and 2.1_m, respectively. The 1.0 and 2.1_m-2219Al retained fine-grained microstructure during annealing treatment at temperatures up to 400 and 425 ∘C, respectively, above which abnormal grain growth occurred. Super plasticity was observed in 1.0_m-2219Al within the medium temperature range of 350–425 ∘C and a maximum ductility of 450% was obtained at 400 ∘C and $3\times10-4$ s−1. Increasing the grain size from 1.0 to 2.1_m resulted in a slight increase in the thermal stability, but did not enhance the super plasticity of the FSP 2219Al. The relatively low super-plasticity in the FSP fine-grained 2219Al was attributed to unstable grain structure because the pining Al2Cu particles were easily coarsened at high temperatures. Fine-grained 2219Al alloys with a grain size of 1.0 and 2.1_m were produced via FSP at 400rpm and 100mm/min in water and air, respectively. The fine metastable precipitates in the 2219Al- T6 were significantly coarsened. The 1.0 and 2.1_m-2219Al retained fine grain structure up to 400 and 425 ∘C, respectively. Above 400 and 425 ∘C, abnormal grain growth occurred in the two FSP sample.

- **Mondal *et al.*, (2008)** show the present investigation, the wear behaviour of a creep-resistant AE42 magnesium alloy and its composites reinforced with Saffil short fibres and SiC particles in various combinations is examined in the longitudinal direction i.e., the plane containing random fibre orientation is perpendicular to the steel counter-face. Wear tests are conducted on a pin-on-disc set-up under dry sliding condition having a constant sliding velocity of 0.837 m/s for a constant sliding distance of 2.5km in the load range of 10–40 N. It is observed that the wear rate increases with increase in load for the alloy and the composites, as expected. Wear rate of the composites is lower than the alloy and the hybrid composites exhibit a lower wear rate than the Saffil short fibres reinforced composite at all the loads. Therefore, the partial replacement of Saffil short fibres by an equal volume fraction of SiC particles not only reduces the cost but also improves the wear resistance of the composite. Microstructural investigation of the surface and subsurface of the worn pin and wear debris is carried out to explain the observed results and to understand the wear mechanisms. It is concluded that the presence of SiC particles in the hybrid composites improves the wear resistance because these particles remain intact and retain their load bearing capacity even at the highest load employed, they promote the formation of iron-rich transfer layer and they also delay the fracture of Saffil short fibres to higher loads. Under the experimental conditions used in the present investigation, the dominant wear mechanism is found to be abrasion for the AE42 alloy and its composites. It is accompanied by severe plastic deformation of surface layers in case of alloy and by the fracture of Saffil short fibres as well as the formation of iron-rich transfer layer in case of composites.

- **Mishra *et al.*, (2002)** search based on FSP in which tool material used is HSS having Rc 59 on CNC vertical milling machine having pin length of 1mm. The tool spindle angle of 2.58 was used. Depth of cut is 1.78 ,2.03 and 2.78 mm resp. with tool transverse rate 25.4 mm/min with constant tool rotating rate of 300 rpm was adopted. Al/ SiC surface composites with different volume fractions of particles were fabricated. The volume fraction of SiC particles was estimated to be 27%.The thickness of the surface composite layer ranged from 50 to 200 mm. The SiC particles were uniformly distributed in the aluminium matrix. The surface composites have excellent bonding with the aluminium alloy substrate. The micro hardness of the surface composite reinforced with 27vol.%SiC of 0.7 mm. (plate thickness is 6mm). Finally result obtained as that too small target depth (1.78 mm) was also ineffective to mix SiC particles into aluminium alloy. A target depth of 2.03 mm resulted in incorporation of SiC particles into aluminium matrix with transverse rate 28.4mm/min. By controlling processing parameters, surface Al/SiC composite layers of 50/200 mm with well-distributed particles and very good bonding with aluminium substrate were generated. So from this research it is clear that friction stir process is well suited for fabrication of composite because FSP tool can flow the material very well, result excellent mixing of matrix material in base material hence composite fabrication can be achieved as desired. Overall study show that friction stir process best suited technique for fabrication of composites die to better material mixing at efficient heating of workpiece.

- **Mironov *et al.*, (2010)** research based on FSP in which he show that Microstructure and texture evolution in the near-surface layer during friction stir processing (FSP) of AZ31 magnesium alloy was studied. Material flow was found to be a very complex process consisting of several stages. The material in front of the friction stir tool was first deformed by the rotating shoulder. Then, approaching the tool, it experienced a secondary deformation caused by the rotating pin, and finally, behind the tool, it again underwent a tertiary deformation induced by the shoulder which clears that tool is responsible for better material flow rate during processing.

- **Ming *et al.*, (2005)** studied high-pressure die cast magnesium alloy AM50 is currently used extensively in large and complex shaped thin-wall automotive components. For further expansion of the alloy usage in automobiles, novel manufacturing processes need to be developed. In this study, squeeze casting of AM50 alloy with a relatively thick cross section was carried out using a hydraulic press with an applied pressure of 70 MPa. Microstructure and mechanical properties of the squeeze cast AM50 with a cross-section thickness of 10 mm were characterized in comparison with the die cast counterpart. The squeeze cast AM50 alloy exhibits virtually no porosity in the microstructure as evaluated by both optical microscopy and the density measurement technique. The results of tensile testing indicate the improved tensile properties, specifically ultimate tensile strength and elongation, for the squeeze cast samples over the conventional high-pressure die cast parts. The analysis of tensile behaviour show that the strain-hardening rate during the plastic deformation of the squeeze cast specimens is constantly higher than that of the die cast specimens. The scanning electron microscopy fractography evidently reveals the ductile fracture features of the squeeze cast alloy AM50. Squeeze casting as a novel manufacturing process is capable of eliminating porosity in magnesium alloy AM50 with a relatively thick cross section compared with high-pressure die casting.

- **Olivier *et al.*, (2009)** studied material flow during friction stir welding is very complex and not fully understood. Most of studies in literature used threaded pins since most industrial applications currently use threaded pins. However, initially threaded tools may become unthreaded because of the tool wear when used for high melting point alloys or reinforced aluminium alloys. In this study, FSW experiments were performed using two different pin profiles. Both pins are unthreaded but have or do not have flat faces. The primary goal is to analyse the flow when unthreaded pins are used to weld thin plates. Cross-sections and longitudinal sections of welds were observed with and without the use of material marker to investigate the material flow. Material flow with unthreaded pin was found to have the same features as material flow using classical threaded pins: material is deposited in the advancing side (AS) in the upper part of the weld and in the retreating side (RS) in the lower part of the weld; a rotating layer appears around the tool. However, the analysis revealed a too low vertical motion towards the bottom of the weld, attributed to the lack of threads. The product of the plunge force and the rotational speed was found to affect the size of the shoulder dominated zone.

- **Ranjit *et al.*, (2011)** studied aluminium based in situ composites has many advantages over their conventional counterparts. However, a major problem in such composites is the segregation of in situ formed particles at the grain boundaries. In this study, it has been shown for the first time that friction stir processing (FSP) can be used effectively to homogenise the particle distribution in Al based in situ composites A single pass of FSP was enough to break the particle segregation from the grain boundaries and improve the distribution. Two passes of FSP resulted in complete homogenization and elimination of casting defects. The grain size was also refined after each FSP pass. This led to significant improvement in the mechanical properties. The novel feature of the composite is that while the strength and hardness improved substantially after FSP, the ductility was not compromised. Segregation of the in situ formed reinforcement particles is a major difficulty in Al based in situ composites. It is shown that friction stir processing (FSP) can be used effectively to homogenise the particle distribution in Al–TiC in situ composites. A single pass of FSP was enough to break the particle segregation from the grain boundaries and improve the distribution. Two passes of FSP resulted in complete homogenization and elimination of casting defects. The grain size was also refined substantially after each FSP pass. The grain size after second pass was finer. The mechanical properties improved significantly after FSP due to improvement in the microstructure. Although the strength and hardness both increases

- **Sato *et al.*, (2005)** studied the principle and advantages of multi-pass friction stir processing (FSP) for the production of a highly formable Mg alloy, and some convincing experimental results are reported in this paper. FSP is a solid state processing technique which involves plunging and traversing a cylindrical rotating FSP tool through the material. FSP achieved grain refinement and homogenization of the as-cast microstructure in Mg alloy AZ91D. Multi-pass FSP produced a fine homogeneous microstructure having a grain size of 2.7 μm throughout the plate. The plate containing this Friction Stir Processed microstructure exhibited fracture limit major strains six times larger than the diecast plate in the fracture limit diagram (FLD). The present study shows that multi-pass FSP is an efficient production method for a large-scale plate of a highly formable Mg alloy.

- **T. McNelley et al., (2007)** studied the current understanding of restoration (i.e., softening) by recovery and recrystallization during thermo mechanical processing of metallic materials has been summarized in several recent monographs and reviews Abstract—Restoration models for hot working of metals and alloys are reviewed in the context of their applicability to friction stir welding (FSW) and friction stir processing (FSP). Two of these models are used to interpret microstructure and micro texture data for two aluminium alloys subjected to FSP. The need for further experiments and model extensions to accommodate the transients and steep gradients in the strain, strain rate and temperature experienced by materials during FSW and FSP are discussed.

- **T. Freeney et al., (2009)** studied single-pass friction stir processing (FSP) was used to increase the mechanical properties of a cast Mg-Zn-Zr-rare earth (RE) alloy, Elektron 21. A fine grain size was achieved through intense plastic deformation and the control of heat input during processing. The effects of processing and heat treatment on the mechanical and microstructural properties were evaluated. An aging treatment of 16 hours at 200 _C resulted in a 0.2 pct proof stress of 275 MPa in the FSP material, a 61 pct improvement over the cast+T6 condition.

- **Verma (1995)** studied the grain growth during an isothermal treatment at a solutionizing temperature of 540~ has been studied in a composite containing 6061 aluminium alloy matrix with Al2O3 particles. The grain growth law is generally applicable to the composites containing 0.10, 0.15, and 0.20 volume fraction of the Al2O3 particles (VFAP). It has been observed that the grain growth process involves the disintegration of the agglomerated particles first and then particles coalesce at longer solutionizing times in the composite containing 0.20 VFAP. The process of coalescence has not been observed up to a heating time of 20 hours at this temperature in the composites containing 0.10 and 0.15 VFAP. The transmission electron microscopy (TEM) study indicates the generation of a large number of dislocations in both the matrix and the area adjacent to the particles. The dislocation densities at these two locations in the composites increase with an increase in VFAP and the particle size. The micro hardness measurements confirm the microstructural observations, and the hardness values for the composite and the matrix appear to be more sensitive to the particle distribution and the particle size compared to the grain size.

- **Wanchuck et al., (2007)** studied the influence of the stirring pin and pressing tool shoulder on the microstructural softening during friction-stir processing (FSP) and subsequent natural aging behaviour was investigated for a 6061-T6 aluminium alloy. The evolution of hardness profiles in various characteristic regions of the FSP plates was investigated as a function of time from 4 to 5760 hours after the FSP through the thickness of the plates and correlated to the microstructure and residual strain profiles measured by a neutron-diffraction technique. The results show that the microstructural softening and the natural aging observed in the dynamic recrystallized zone and thermo mechanically affected zone are mainly caused by the frictional heating from the tool shoulder, resulting in dissolution and precipitation of strengthening precipitates. On the other hand, the softening in the heat-affected zone is due to the dissolution/growth of the precipitates and is not followed by the natural aging under the current processing condition. The kinetics of the natural aging behaviour is also discussed. Two different FSP specimens were prepared to investigate the effects of the stirring pin and tool shoulder on the microstructural softening and subsequent natural aging in 6061-T6 Al alloy: (Case 1) a plate processed using both tool shoulder and stirring pin and (Case 2) a plate processed using only the tool shoulder.

- **Y. Mazaheri, *et al.*, (2011)** studied that A356/Al2O3 surface nano composite was produced by friction stir processing (FSP) method. X-ray diffractometery, optical and scanning electron microscopy, micro hardness and nano indentation tests were used to characterize the samples. The results indicated that the uniform distribution of Al2O3 particles in A356 matrix by FSP process can improve the mechanical properties of specimens. The hardness and elastic modulus of the as-received A356, the sample treated by the FSP without Al2O3 particles, surface micro and nano composite specimens were about 75 Hv and 74 GPa, 69 Hv and 73 GPa, 90 Hv and 81 GPa, 110 Hv and 86 GPa, respectively. From the experiments and analyses performed, some conclusions can be drawn:

(1) The microstructural study of A356/Al2O3 surface composite layers fabricated by FSP indicated that Al2O3 particles were well distributed in the Al matrix, and good bonding with the Al matrix was generated.

(2) The FSP with Al2O3 particles obviously increased the micro hardness of the substrates. The micro hardness values for A356–_Al2O3 and A356–nAl2O3 surface composite were about 90 and 110 Hv, respectively, while that of the sample treated by the FSP without Al2O3 particles and the as-received A356 were about 67 and 80 Hv, respectively.

(3) The results obtained from nano indentation technique showed better micro hardness and elastic modulus values for A356/Al2O3 surface composites in comparison with as-received A356 and Friction Stir Processed sample (no Al2O3).

(4) The better mechanical properties of A356/Al2O3 surface nano composite can be attributed to the presence of nano sized Al2O3 particles, which contribute significantly to the strength through the Orowan mechanism.

- **Yazdipoura *et al.*, (2009)** develop analytical model to study the effect of cooling rate on the final grain size of stirred zone of the Al5083 subjected to friction stir processing. The effect of cooling rate on the grain size of the stirred zone was investigated experimentally. A new microstructural evolution model was also suggested illustrating the mechanisms contributed in refining the microstructure. A new mechanism termed meta-dynamic recovery (MDRV) is introduced here in this regard. The simulation results also show that the rapid cooling rate resulted in superior mechanical properties through refining the microstructure of the stirred zone. However, decreasing the rotational speed and increasing the traverse speed of pin can also decrease the grain size of stirred zone. The results also show that while heat input to the stir zone may control the _nuclei size and _nucleation mechanisms, cooling rate may considerably affect the _grain growth step.

- **Y. Morisada *et al.*, (2008)** studied the microstructural control of tool steel (SKD11) by laser melting and friction stir processing (FSP). A nano meter-sized microstructure consisting of fine M7C3 carbide (particle size: 100 nm) and a matrix (grain size: 200 nm) was successfully fabricated by the FSP on the laser treated SKD11. The nanostructured SKD11 had an extremely high hardness of about 900HV. The matrix grains and carbide particles of the SKD11were significantly refined by the laser melting and the FSP. The microstructure and micro hardness were evaluated by observations of the grain size and phase of the matrix, and the size and dispersion of the carbide particles.

- **Zahmatkesh *et al.*, (2010)** studied, a new processing technique, friction stir processing (FSP) was applied to Al2024-T4 as a means to enhance the near-surface material properties. Samples were subjected to FSP using a constant tool rotating rate of 800 rpm and travel speed of 25 mm/min with a tool tilt angle of 3degree. Microstructural evolution and tribological behaviour of friction stir processed (FSP) Al2024-T4 were investigated. Microstructural characteristics of the samples were investigated by optical microscopy (OM) and scanning electron microscopy (SEM). Evaluations of mechanical properties include micro-hardness and wear resistance. Dry sliding wear tests were applied using a reciprocating wear test. The results showed that FSP was beneficial concerning improvement of hardness and wear resistance. FSP reduced friction coefficient by approximately 30% and wear rate by an order of magnitude. From this investigation, the following important conclusions are derived The NZ exhibited homogeneous and fine equi axed grains with average size of _4 lm. The maximum hardness was achieved in NZ (_110 Hv). FSP was found to be beneficial in improving wear resistance. in this region.

2.2 Research Gaps

- Complex shaped parts cannot process like circular, elliptical etc.
- There are large number of variables such as different speed and feed, tool dimensions for different which are difficult to select because there are no proper standards and formulation so that these variables can be efficiently select for different kind of materials..
- When process finished exit holes remains on surface of work piece so no techniques formed which immediately filled hole before work piece get cooled.
- The microstructure and mechanical properties of the processed zone can be accurately controlled by optimizing the circular tool design during FSP of aluminium. But it shows the results of only by circular tool profiles, now question arises that if tool profile design changed then what is effect on microstructure and properties of AL 6063 in friction stir process.

2.3 Methodology

- Firstly aluminium plates cut in equal dimensions of 150*100*6mm.
- Four tool profiles were used which made from HSS (high speed steel) having 55Rc hardness
- Each plate is to be process by each tool profile under CNC vertical milling machine (CNC VMC) with keep all parameter constant e.g. speed and feed, so that proper results can be obtained.

- Now after completion of FSP process, test to be performed in which micro hardness microstructure and mechanical were evaluated. Finally comparison of result which obtained through characterization techniques.

CHAPTER 3
EXPERIMENTATION

Friction stir processing (FSP) was carried out using a CNC vertical milling machine set up fitted with a specially designed FSP tool which stirred in vertical axis or perpendicular to workpiece. The base metal (i.e. Al 6063) was fixed in a specially designed fixture that was held on the bed of milling machine. The FSP tool was rotated on the surface of base material. The processed samples was characterized for optical microscopy, micro hardness and impact test. The results were also obtained for base metal (6063Al) for comparison. The worn tracks of the samples were investigated under scanning electron microscope. The details of various equipments, techniques and procedures used in the study are as following.

3.1. Process parameters and procedure

All the parameters kept constant during process for all four tool profile as given table 3.1. Table 3.2 show plate dimensions used for FSP with each tool pin profile.

Table 3.1: process parameters

Tool profile	Process parameter
Pentagonal tool pin profile	Spindle speed=1000 feed rate=19mm/min
Threaded tool pin profile	Spindle speed=1000 feed rate=19mm/min
Circular tool pin profile	Spindle speed=1000 feed rate=19mm/min
Square tool pin profile	Spindle speed=1000 feed rate=19mm/min

Table 3.2 Specimen Dimensions

PROFILE	PLATE DIMENSIONS
pentagonal tool pin profile	L=150mm, w=100mm, t=6mm
square tool pin profile	L=150mm, w=100mm, t=6mm
threaded tool pin profile	L=150mm, w=100mm, t=6mm
circular tool pin profile	L=150mm, w=100mm, t=6mm

3.2 Material

3.2.1 Base material

In this study the FSP was performed on plate made of 6063 aluminium alloy. Chemical compositions is given in table 3.3

Table 3.3: chemical composition of AL 6063

Element	Weight t%
Cu	0.082
Mg	0.619
Si	0.600
Fe	0.350
Manganese	0.044
Ni	0.006
Zn	0.061
Pb	0.046
Tn	0.024
Titanium	0.015
Cr	0.007
Al	97.94

This alloy (AL6063) is known for architecture material performance in extreme environments. Al 6063 is mostly use in domestic applications like in window frames and doors, automobile engineering. Al 6063 is highly resistant to attack by both seawater and industrial chemical environments, and it also retains exceptional strength after welding. It possesses the highest strength amongst the non-heat treatable alloy.

Aluminium 6063
Length=150mm
Width =100mm
Thickness=6mm

Fig. 3.1 Base plate AL 6063

3.3 Equipments
3.3.1 CNC vertical milling machine

CNC vertical milling machine used for friction stir processing of aluminium 6063 sample plates. After selection of the entire machine parameters next step is experimentation. So for FSP CNC vertical milling machine selected which having pneumatic tool holder. In the CNC vertical milling the spindle axis is vertically oriented. Milling cutters are held in the spindle and rotate on its axis, position of tool is also vertical which is held in pneumatic chuck. The spindle can generally be extended (or the table can be raised/lowered, giving the same effect), allowing plunge cuts and drilling. There are two subcategories of vertical mills: the bed mill and the turret mill. The Operating system of such machines is a closed loop system and functions on feedback. These machines have developed from the basic NC (NUMERIC CONTROL) machines. A computerized form of NC machines is known as CNC machines. A set of instructions (called a program) is used to guide the machine for desired operations.

Fig. 3.2: CNC Vertical milling machine

Also, various other codes are used. A CNC machine is operated by a single operator called a programmer. This machine is capable of performing various operations automatically and economically.

Table 3.4: Specifications of CNC vertical milling machine

Travel	Metric
X-axis	762 mm
Y-axis	406 mm
Z-axis	508 mm
Table length	914 mm
Table width	356 mm
T-slot	16
No of T slot	3
Spindle maximum rating	22.44 kw
Spindle maximum speed	8100 rpm
Maximum torque	122 N-m
Bearing lubrication	Air/oil injection
Cooling	Air
Motor capacity	20
Max tool diameter	89 mm
Air required	1113 l/min
Coolant capacity	208 L

While reading this article, along with collecting other information about milling machines, it is crucial for one to understand the computerized form of such machines. In tool holding mechanism, improvement on CAT Tooling is BT Tooling, which looks similar and can easily be confused with CAT tooling. Like CAT Tooling, BT Tooling comes in a range of sizes and uses the same body taper. However, BT tooling is symmetrical about the spindle axis, which CAT tooling is not. This gives BT tooling greater stability and balance at high speeds. One other subtle difference between these two tool holders is the thread used to hold the pull stud. CAT Tooling is all Imperial thread and BT Tooling is all Metric thread. Note that this affects the pull stud only, it does not affect the tool that they can hold, both types of tooling are sold to accept both Imperial and metric sized tools.

3.3.2 Tool holder

Basic function of tool holder is to holds the tool in desired axis in CNC VMC machine. In experimentations pneumatic tool was used. An improvement on mechanical Tooling is pneumatic Tooling, which looks similar and can easily be confused with pneumatic tooling. Like mechanical tooling, pneumatic tooling comes in a range of sizes and uses the same body taper. However, tooling is symmetrical about the spindle axis. This gives tooling greater stability and balance at high speeds. In these experimentations pneumatic tooling used which save much time than mechanical tooling during changing of tools. Following diagram shows simple tool holder used for holding the cylindrical tools.

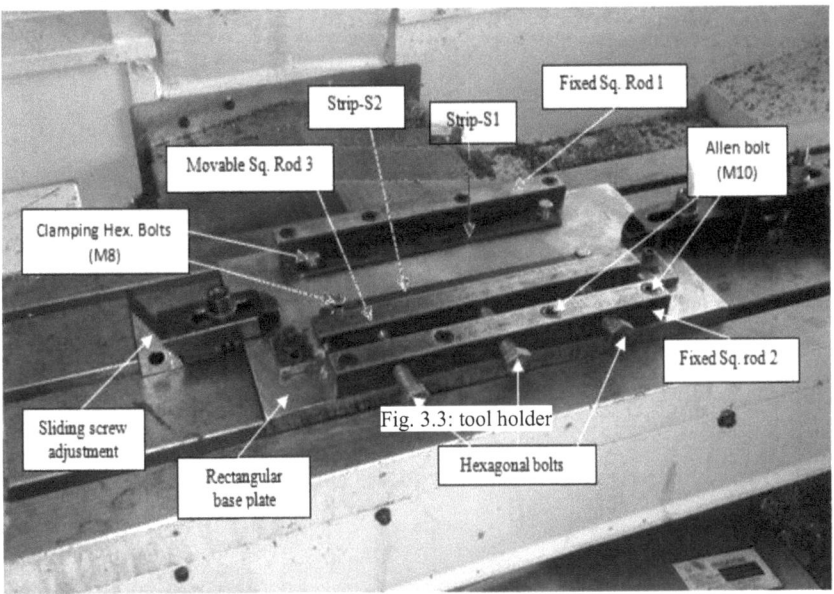

Fig: 3.3. Fixture used in experimentations

3.3.3 Fixture

The fixture for holding the base plate, while carrying out FSP was designed in house and fabricated at Dhiman Industries, Bathinda. The fixture consisted of a rectangular base of dimensions 400mm x 200mm x 20mm. Three numbers of square rods having X-section (25mm x25mm) and length 300mm were machined to an accuracy of 5µm. Out of these square rods, two rods were drilled with counter sunk holes (4 Nos.) to adjust bolts of size M10.

Fig: 3.4. Fixture hold Al plate on bed of CNC vertical milling machine

These two rods were fixed at the ends of rectangular base plate. The Sq. rod 2 was also consisted of 3-addition drilled holed consisting of M12 internal threads to accommodate hexagonal bolts of size M12. The third square rod was placed between Sq. rods 1 and 2 and was movable with the help of hexagonal bolts tightened to square rod 2. The base plate to be friction stir process was held tight between square rods 1 and 3. The base plate was also tightened to the rectangular base plate with the help of 2 MS strips (S1 and S2) each screwed to the rectangular base plate with the help of hexagonal bolts (M8). Fixture show excellent clamping of 150*100*6mm dimension plate as shown in following diagrams. First diagram show fixture component an second diagram show clamping of plate. This fixture can holds the rectangular and square plate specimens very well and avoid the any movement of plate with respect to machine table.

3.4 FSP Tools

Following four types of tool pin profile was used during the experimentations
- Circular tool pin profile

- Pentagonal profile
- Threaded tool profile
- Square tool profile

• *Circular tool pin profile*

Mostly used tool pin profile is circular tool pin profile as shown in following diagram. It simpler in design and can be machined easily on centre lathe. Tool pin length varied according to the thickness of work piece, it should be ½ thickness of work piece and shoulder should touch with surface of material to be processed as shown in first diagram. Now using the working material get deformed and melts during single pass of tool and after pass complete pass material solidified during process with circular tool profile because to pin is round less power is required and also less wear tear of pin than that of pentagonal tool profile.

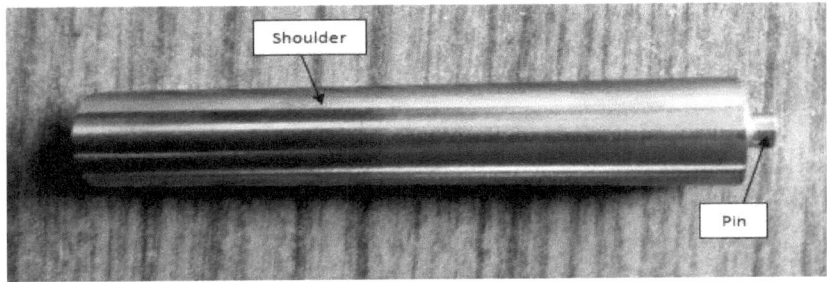

Fig: 3.5. FSP tool made from HSS

Tool specification
Material=HSS (high speed steel)
Shoulder length=60mm
Shoulder diameter=10mm
Pin length=4mm
Pin diameter=3mm
Rockwell hardness=59

• *Square tool pin profile*

With increasing experience and some improvement in understanding of material flow, the tool geometry has evolved significantly. Complex features have been added to alter material flow, mixing and reduce process load. Another tool profile is square tool pin profile which having square base of pin tip, remaining dimensions same as circular profile so we can say that square tool profile is modified form of circular tool profile.

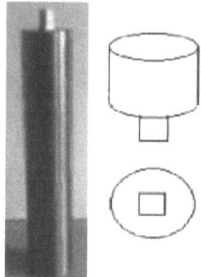

Fig: 3.6. Square tool used for FSP

Square tool profile has four sharp edges. If four edges of tool profile replace with six edges a new profile will formed known as pentagonal tool profile. Tool dimensions are:

Dimensions of square tool profile, Shoulder length=60mm, Pin length=4mm, Shoulder diameter=10mm, Square pin = 4mm square

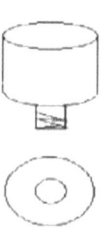   **Threaded Pin**

   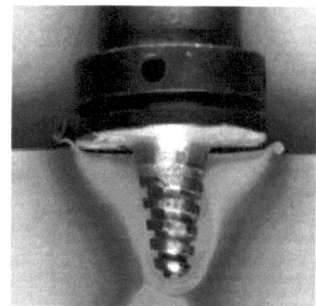

Fig: 3.7. Drawing of the FSP tool (R.S. Mishra *et al.*, 2005)

- *Threaded tool profiles*

Another tool profile is threaded tool pin profile that pins tools are shaped as a threading that displaces less material than a cylindrical tool of the same root diameter. These tool pin are mostly used in composite fabrication due to good material mixing by threaded profile Typically, reduces displaced volume by about 60%, Tool dimensions are as, Shoulder length=60mm, Pin length=4mm, Shoulder diameter=10mm Pin = clockwise threads on diameter of 3 and length of 4mm.

- *Pentagonal tool profile*

Pentagonal tool profile same as square tool profile except that there are 6 sharp edges on pin circumference. Diagram show front view and top view pentagonal tool profile. Main difference is that circumference volume is more and sharp edges are more responsible for material flow rate but wear tear is more than other profiles. Tool dimensions, Shoulder length=60mm, Pin length=4mm, Shoulder diameter=10mm Tool pin =pentagonal shape at angle of 60 degree in circle of 3mm dia. (six edges)

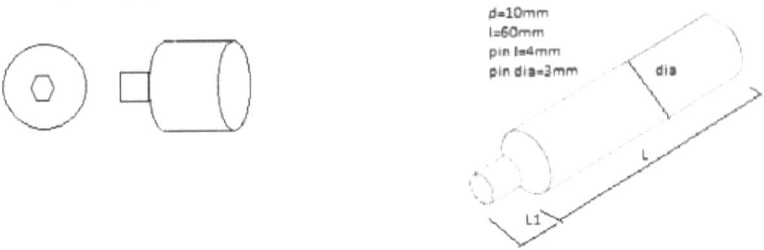

Fig: 3.8. Pentagonal tool pin profile

3.5 FSP procedure

Friction stir processing of the base plate (6063 Al) was performed on a CNC vertical milling machine, (made by HAAS automations INDIA) a using a FSP tool, as described earlier. Before FSP, the plate having dimension 150*100*6mm, was gripped in the fixture, as described earlier. The fixture containing duly gripped base plate was placed and tightened with the bed of CNC vertical milling machine. Commands are given to machine with spindle speed of 1000rpm and feed rate 19mm/min. Tool depth given 3mm in work piece surface.

Fig. 3.9: FSP

Speed FSP tool: 1000 rpm, Feed rate: 19mm/min, Depth of tool penetration in plate: 3mm (1/2 thickness of plate), Angle of FSP tool: Nil (w.r.t normal to plate surface). Tool firstly start rotating in air and then plunged into workpiece where tool start stirring at 1000rpm (nugget zone). All the procedure repeated by each four profiles i.e. pentagonal, rectangular, threaded and circular tool pin profile at same constant parameters. Tool holder holds tool by pneumatic action, so it was very easy to change tool again and again during operation. Four plates were processed by four tool pin profiles with one each. Results obtained are discussed in next chapter.

3.6 Characterization and testing of samples

In order to investigate the effect of tool pin profiles of 6063Al, the FSP samples were subjected to microstructure and mechanical tests (micro hardness, hardness and impact test). The microstructures of the samples were observed through optical microscopy while the work surfaces were examined under scanning electron microscope (SEM). The details of testing and characterization carried out in the study are described in the following section.

3.6.1 Microstructural examination of al 6063 samples

The microstructures of the samples were observed under inverted optical microscope having PC interface and Measurement software as shown in figure. Microstructure of all four plates tested under optical microscope at magnification rate 500xx.microrstructure testing conduct for checking internal changes in material or say effect of tool profiles on microstructure. Four microstructure test were conducted which gives output result different for different tool profiles. The cross section of the FSP aluminium 6063 plate was cut in the middle of the plate length (LD) for the micro structural characterization .For the proper visualizations the cross-sectional sample was cold mounted, ground, polished, and etched with the Keller's reagent (at room temperature for optical microscopy at the rate of 100 magnifications). All the four specimens are cut in to the small size circular foils so that these can be placed accurately on optical microscopy machine. In addition, transmission electron microscopy observations were performed to investigate precipitates. The disc specimens were prepared by grinding of specimen foil, and electro polishing using nitric acid solution in methanol with 25 voltages for 20 seconds at) 30°C. The four specimens prepared for viewing their microstructure. Firstly the friction stir processed samples were cut into four small sizes along the longitudinal as well as transverse sections. The cutting was performed with the help of hand hacksaw so as to avoid heating of the specimens, which might have cause microstructural distortions. (Four specimens are shown in fig. 3.12) The cut pieces were mounted inside thermosetting plastic. The samples were placed inside a hollow pipe (dia. =50mm, length=20mm) with pressed surface downward. Finally finished specimen were tested under optical microscope as shown.

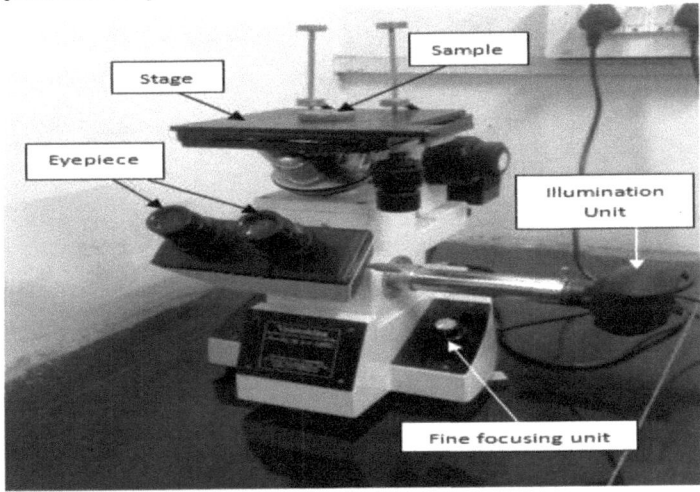

Fig. 3.10: optical microscope

All four specimens grinded manually, for purpose to examine microstructure clearly and precisely.

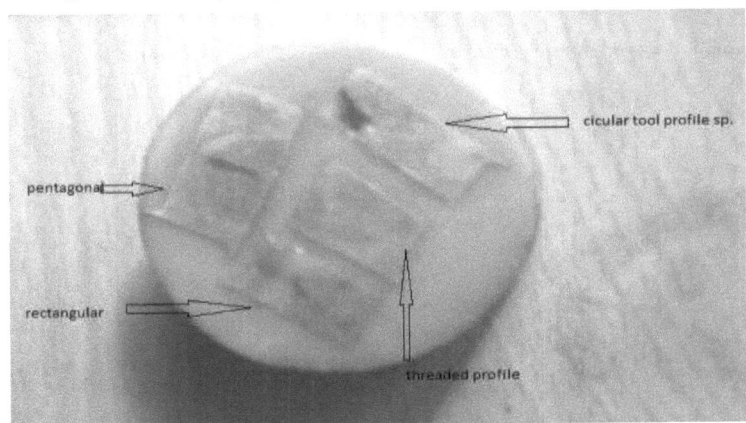

Fig. 3.11: Al sample mounted on plastic foil for microstructure examination

The mounted samples were first ground on the flat belt grinder followed by manual grinding on successive grades of emery papers starting from 250 grits to 400, 800 coarse to fine grits of 1600 and 2000. The polishing was done until mirror finished and crack free surface was obtained. The final polishing was done on buffing machine by using diamond paste and continuously running water.

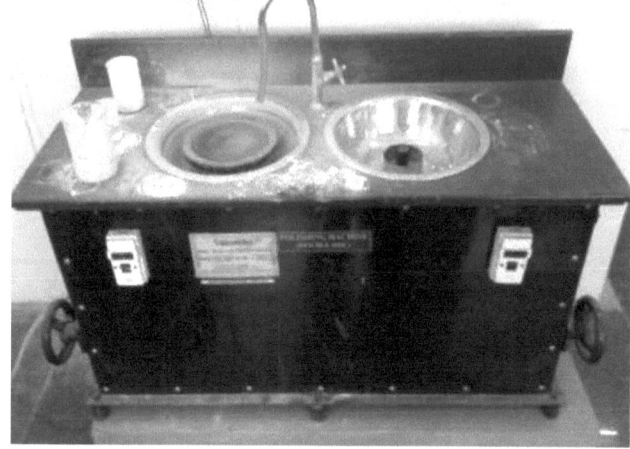

Fig. 3.12: Buffing machine.

3.6.2 Micro hardness measurement

Micro hardness testing conducted here to test hardness of thin layer of specimen. So in this testing there is only small indentations (1mm) which test the hardness of processed layer. Rockwell micro hardness (Hv) was measured along the polished cross section using 100 g of the applied load. There are four micro harden test on aluminium plates

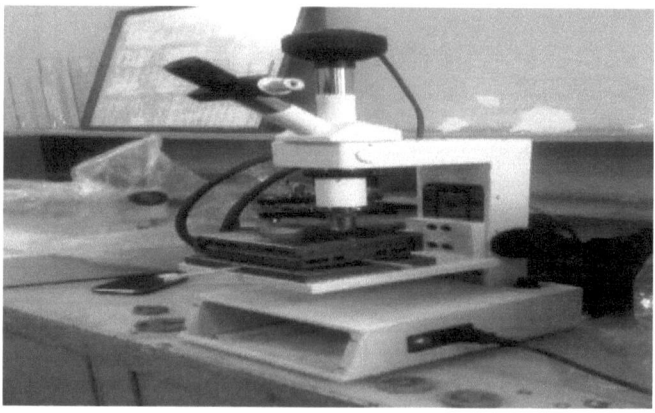

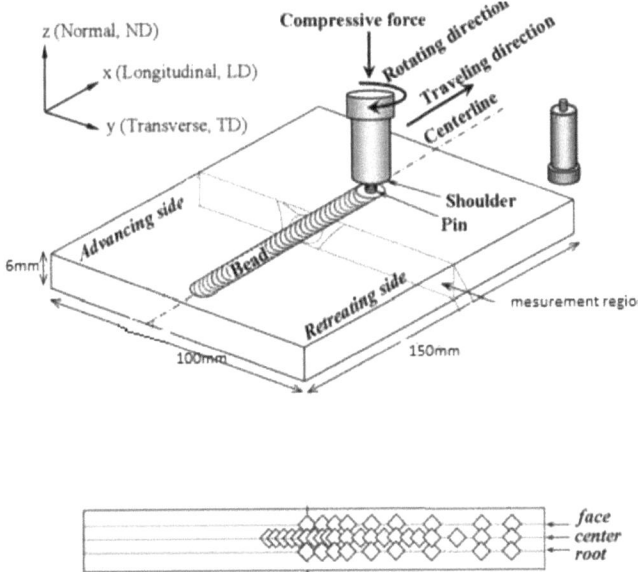

Fig. 3.13: micro hardness measurement machine and mechanism

A total of three lines were measured with horizontal spacing and through the thickness of 3mm of the plate along the middle of the plate, micro hardness test conducted on homogeneous fine grained microstructure. Four tool profiles show different result in case of micro hardness. The size of indention was measured with the help of micrometer and eyepiece (having magnification of the order 10X) fitted on the hardness tester. The impression was created with the help of indenter (Pyramid shape) by applying loads, ranging from 20gms to 200gms

3.6.3 Izod impact strength testing

Impact tests provide information on the resistance of a material to sudden fracture where a sharp stress riser or flaw is present. In addition to providing information not available from any other simple mechanical test, these tests are quick and inexpensive. The data obtained from such impact tests is frequently employed for engineering purposes. Impact testing on FSP aluminium 6063 conducted because Al 6063 is domestically used material (used in window, door frame and architecture engineering) where impact load always act on material. So it is necessary to find out impact strength of material before and after FSP. In present study izod test conducted on al 6063 specimens as shown in following diagrams.

Fig. 3.14: Izod impact testing machine

Various standard impact tests are widely employed in which notched specimens are broken by a swinging pendulum. In this testing IZOD impact testing used to find out the impact strength.

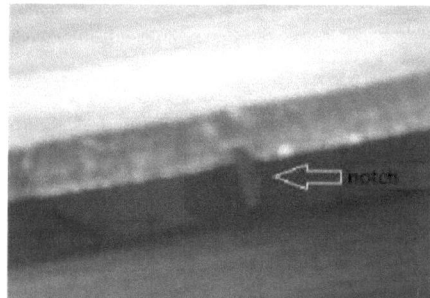

Al 6063 sample prepared for impact testing   fig show notch on sample

Fig. 3.15: Al 6063 sample prepared for impact testing

Four this purpose specimen cut into desired shape by hand hacksaw and notch cut up to 2mm in processing zone for weaken the processing zone area so that impact strength of processed area can be measured as shown in above diagram.

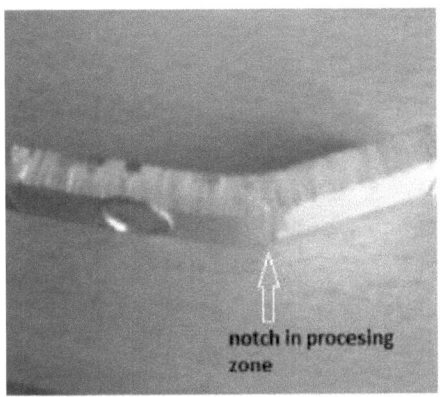 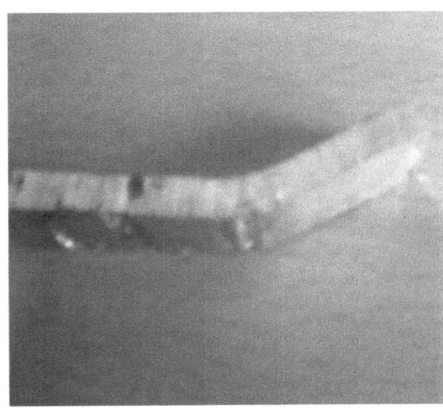

Specimen fracture after izod impact testing

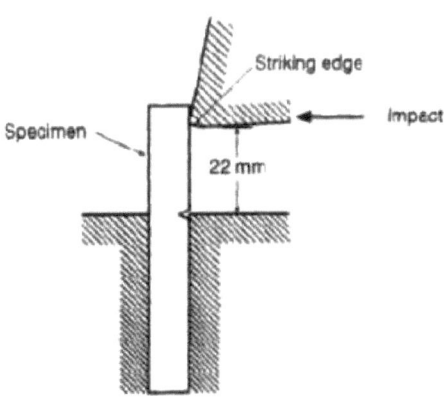 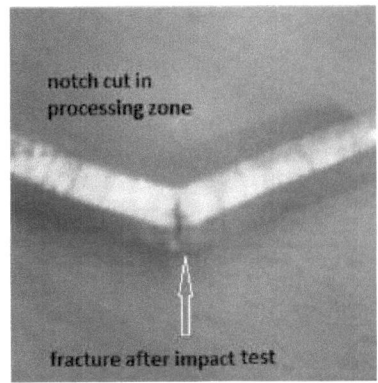

Fig. 3.16: striking position and specimen fracture after izod test

Above diagrams show standard dimensions of specimen for hold on vice and R.H.S. diagram show specimen after impact testing. Results of test for each profile are discussed in next chapter.

3.6.4 Rockwell hardness testing

To check the Rockwell hardness processed specimens grinded into 10*10 mm sizes. As shown in following diagrams internal (red) scale used for hardness measurement.1kg of load applied on plate surface after which different results were obtained which are discussed in next chapter. Pyramid shape indenter was used for testing size of 1/6‖ ball.

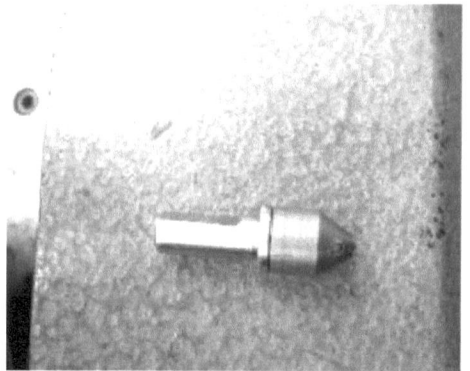 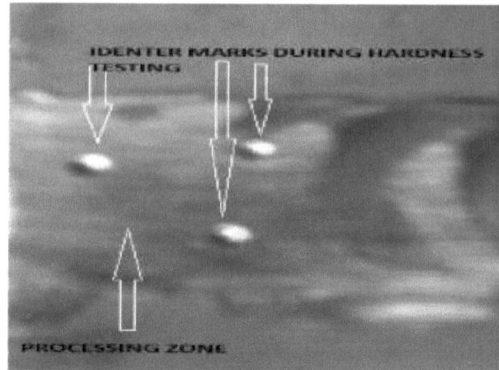

Fig. 3.17: Pyramid shape indenter and indenter marks in processed zone

Fig. 3.18: Rockwell hardness testing machine

CHAPTER 4

RESULTS AND DISCUSSION

4.1 Microstructure results

Fig. 4.1: microstructure of AL base plate

Microstructural characterizations include optical microscopy, scanning electron microscopy (SEM), energy-dispersive/ X-ray spectroscopy (EDS or EDX). Optical microscopy of aluminium base plate and processed plate was used to examine the microstructure which include grain size and transition zone in the sample. Identification of the operating wear mechanisms was ascertained through the detailed analyses of worn tracks by means of Scanning Electron Microscopy (SEM) and Energy Dispersive X-ray analysis (EDX).

Samples for metallographic examination were cut from the surface of base and Friction Stir Processed samples. These samples were wet grounded using various grades of emery paper and polished to attain mirror finish by using buffing machine. The etching of the samples was done with the reagent 0.5 ml HF (40%) diluted with 99.5 ml distilled H_2O before viewing the microstructure under optical microscope. Optical micrographs of Al-6063 base alloy and Friction Stir Processed alloy samples were viewed at magnification of 100 and are depicted in Fig. 4.1 to fig. 4.5. the Friction Stir Processed alloy sample. The fine grain size observed in Friction Stir Processed aluminium sample is responsible for its high hardness than the base metal. The grain refinement in the processed sample may be due to the occurrence of dynamic recrystallization phenomenon as a result of disruptive mechanical action of the tool pin profile provided in the FSP tool. Base plate microstructure consists of casting impurities and also having non homogeneous structure. So after examination it was observed that as received base plate having coarse grained microstructure with surface impurities.

4.1.1 Pentagonal tool profile result in microstructure of aluminium 6063

After processing with pentagonal tool pin profile fine grain microstructure results obtained. Principle behind this is that there are edges on profile of pentagonal tool profile which are responsible for efficient (more and more) material flow and heat generation(effective crystallizations in nugget zone) during the friction stir processing, so it is reason that why microstructure become more uniform and crack free after processing. But main disadvantages is more wear tear of edges of pentagonal tool profile than circular profile.

Fig. 4.2: Processed zone microstructure for pentagonal tool pin profile

4.1.2 Square tool pin profile microstructure results

On second hand square tool profile microstructure obtain is fine grained but cracks and cavity appear in the processing zone, it means that material flow and heat dissipation is not so efficient as pentagonal tool profile .There is little difference in results of square tool pin profile than that of pentagonal tool profile.

Fig. 4.3: Microscopic image of microstructure of aluminium 6063 sample processed with square tool pin profile

4.1.3 Threaded tool pin profile microstructure results

Threaded pin profile produces average stirring action in nugget zone. Microstructure less homogeneous because it shows more cracks and cavity in processed zone. Discontinuity has been observed in processed zone which increase porosity in microstructure. So, microstructure is having more porosity with homogeneity in microstructure of nugget zone.

Fig. 4.4: Microscopic image of microstructure of aluminium 6063 sample processed with threaded tool pin profile

4.1.4 Circular tool pin profile results

Due to no edges on tool profile material movement was not so good hence small cavities in nugget zone of processed material but microstructure is fine grain and also less porous than base plate as in case of other above three profiles.

Fig. 4.5: Microscopic image of microstructure of aluminium 6063 sample processed with circular tool pin profile

Table.4.1: microstructure results

Tool profile	Parameters	Results
Pentagonal pin profile	Speed=1000rpm.feed=19mm/min	Microstructure obtained fine grained homogeneous. There is no cavity/crack formed in stirring zone. No any discontinuity in processing zone.
Square tool profile	Speed=1000rpm.feed=19mm/min	Microstructure obtained is fine grained and homogeneous but still cracks appears in stirring zone and aluminium silicates still appears in grain boundries.
Threaded tool profile	Speed=1000rpm.feed=19mm/min	Microstructure less homogeneous because it shows more cracks and cavity in processed zone .Discounty have been observed in processed zone which show more porosity in microstructure.
Circular tool profile	Speed=1000rpm.feed=19mm/min	Due to no edges on tool profile, material movement was not so good hence small cavities in microstructure but material become more porous than base plate as in case of other above three profiles
Base plate	Unprocessed al sample	Coarse grained microstructure Non homogeneous poor structure

It is clear that each give fine grained microstructure but different results in cracks and cavities in nugget zone as discussed in above table. All the result of microstructure are shown in following graph which clears in which we observe that pentagonal tool profile gives better results than other three tool profiles. Following diagram shows that pentagonal tool profile gives more superior microstructure results than other profiles.

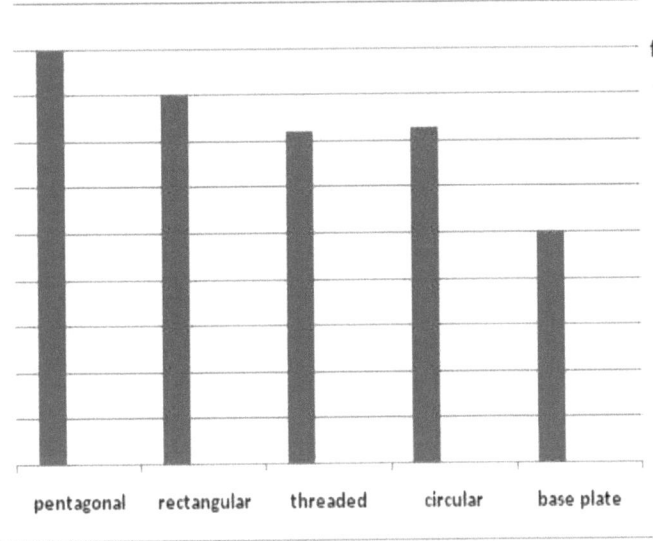

profile results w.r.t. microstructure

Fig. 4.6: microstructure comparison with each tool pin profile processed specimen

4.2 Results of micro hardness

Micro hardness was measured across the top surface as well as across the transverse section of the Friction Stir Processed and base metal samples. Micro hardness profiles measured across the top surface of the sample. The hardness measurements were performed on both sides of the FSP zone centre at an interval of 2 mm. Similarly the micro hardness of the base alloy was also measured. It is observed that hardness of the Friction Stir Processed zone is significantly higher than that observed for the base alloy. The hardness profile shows that the hardness is maximum (51.2HV in case of pentagonal tool pin profile) at the centre of Friction Stir Processed zone (nugget/stir zone).However, the micro hardness of the base alloy remains almost uniform, 47 HV.

It is clear that processed specimen micro hardness increases than that of base plate and out of processed samples with each four profile, pentagonal tool pin profile gives maximum results, results show in following table.4.2.To find out influence on mechanical properties microharness test conducted for each specimen and following results were obtained:-

Table.4.2: micro hardness comparison results

TOOL PROFILE	PARAMETERS	RESULTS
Pentagonal pin profile	Speed=1000rpm.feed=19mm/min	Micro hardness=51.2
Square tool profile	Speed=1000rpm.feed=19mm/min	Micro hardness=50.1
Threaded tool profile	Speed=1000rpm.feed=19mm/min	Micro hardness=50
Circular tool profile	Speed=1000rpm.feed=19mm/min	Micro hardness=50.55
Base plate	Not processed	Micro hardness=47

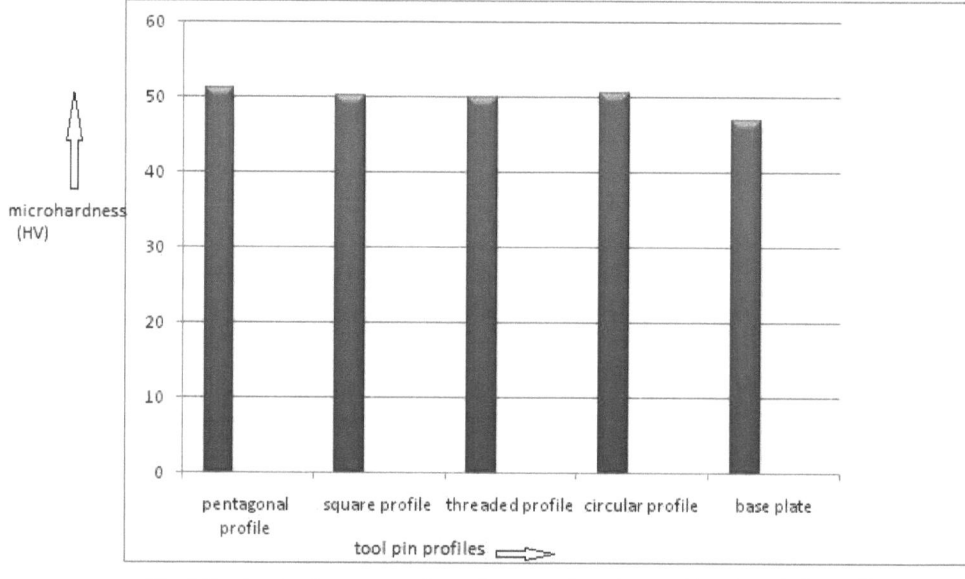

Fig. 4.7: micro hardness comparison of processed specimen with each tool pin profile

Observations and results show that specimen processed with pentagonal tool profile show best results than other three profile (same as microstructure case) .So it is clear that fine grained microstructure without cracks, having more micro hardness (in this research) as pentagonal profile show. So there is direct relation in microstructure and micro hardness in this research. It is clear finally that specimen processed with pentagonal tool pin profile give fine grain microstructure without any crack or cavity in processing zone and maximum micro hardness out of four tool pin profiles. Above diagram also show that there is slightly change in tool pin design so that results i.e. micro hardness also slightly varied w.r.t. tool design.

4.3 Impact strength results

Aluminium specimen both processed and unprocessed were tested for find out impact strength variation in which test show that material bends, deformed and does not completely broken as brittle material (cast iron). Impact strength of processed specimens was measured in processed zone area by providing notch in processed zone (as shown in previous chapter) and base plate also. After results it has been found that processed zone having more impact strength (12.53) than base plate (12 joule). It is clear that there is also increase in impact strength of FSP processed specimen.

Table: 4.3 Impact strength results

Tool profile	Impact load (kg)	Impact strength (in Joule)
Pentagonal pin profile	5	12.53
Square pin profile	5	12.43
Threaded tool profile	5	12.22
Circular tool profile	5	12.11
Base plate	5	12.00

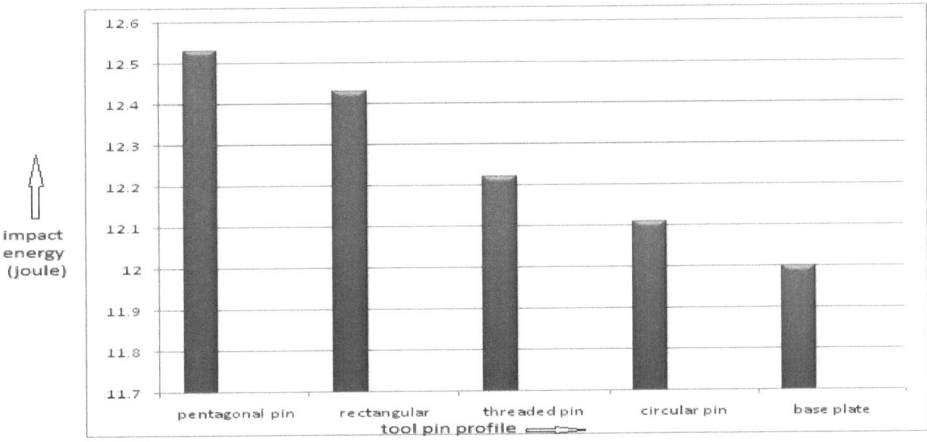

Fig. 4.8: impact strength comparison of FSP processed sample with each tool pin profile

Out of processed specimen, pentagonal tool pin profile specimen show more impact strength as shown in table: 4.3.There is also cleared that material having uniform fine grained microstructure, having more impact strength and mechanical properties as cleared in this research.

After result it is find out that there is slightly variations in impact strength, impact strength slightly increases than base plates. But same pentagonal profile gives maximum results i.e. 12.53 joules.

4.4 Rockwell hardness test results

Friction stir processed sample plates show more hardness than base metal because processed plate having fine grain crystallized homogeneous microstructure which is responsible for more hardness than unprocessed base plate. Rockwell hardness of each processed specimen was measured in centre and sides and mean of three values was evaluated.

It was observed that side of nugget zone sides having slightly more Rockwell hardness than its centre. As shown in table.4.4 out of four tool pin profile pentagonal tool profile processed plate specimen having maximum Rockwell hardness HRB 35.5.

Table.4.4: Rockwell hardness results

Tool pin profile	Parameter(load)	Results(hardness)
Pentagonal tool pin processed specimen	1kg	HRB 35.5
Rectangular tool pin processed specimen	1kg	HRB 32
threaded tool pin processed specimen	1kg	HRB 30
circular tool pin processed specimen	1kg	HRB 29
Base plate specimen	1kg	HRB 24

After impact testing specimen processed with pentagonal tool pin profile show maximum hardness out of three profiles at constant parameter as shown in above table. Maximum hardness is HRB 35.5. of pentagonal tool pin processed specimen reason is that pentagonal tool pin prole specimen having fine grain microstructure without any crack and cavty in nugget zone.There is increase in hardness of processed specimen than base plate in which base plate having hardness of HRB 24.

Other tool pin profile also show better results than base plate as shown in above table. There is also disadvantage of pentagonal tool pin profile more wear tear of sharp edges than other tools, so for mass production these tools are not suitable because after every 2 plate (150*100*6 mm) new tool is required on other hand circular and threaded tool pin profile can be used for ma production applications.

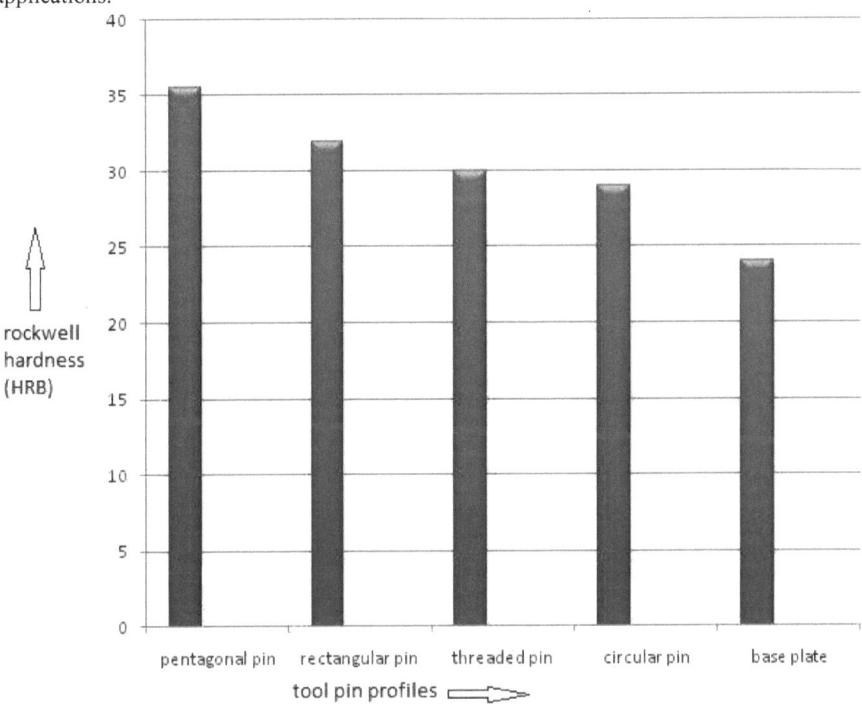

Fig. 4.9: comparison between Rockwell hardness of FSP processed specimen

As results obtained it is find out that there is little increase in Rockwell hardness of processed specimens from base plate. Second thing show that specimen processed with pentagonal tool pin profile having more hardness (as previous cases).

Finally Best results obtained by profiles (in all tests) are as

$$\text{Pentagonal tool pin profile} \\ \downarrow \\ \text{Square tool pin profile} \\ \downarrow \\ \text{Threaded tool pin profile} \\ \downarrow \\ \text{Circular tool pin profile}$$

CHAPTER 5

CONCLUSION AND FUTURE SCOPE

5.1 Conclusions

Following conclusions drawn from present research work

- First of all it is clear that all FSP processed specimen gives fine grain microstructure and maximum mechanical properties than Al6063 base plate (unprocessed).
- Secondly, effect of each pin profile examined by conducting the test on specimens, processed by each profile. Results show that pentagonal tool pin profile processed plate gives best result in every test. Reason behind that pentagonal tool pin profile having sharp edges with which are responsible for more stirring action and crystallizations in nugget zone. That by fine grained microstructure without an cavity in nugget zone, obtained after processing with pentagonal tool pin profile (as discussed in microstructure results)
- Maximum micro hardness obtained by pentagonal tool profile (51.2 Hv) due to fine grained microstructure which shows direct relation between microstructure and micro hardness.
- Also there is slightly increase in impact strength and Rockwell hardness of processed plates than base plate.

5.2 Scope for Future Work

- In future this research help in selection of tool pin profile to get maximum desired results.(as it is clear that pentagonal profile show maximum results)
- Further research can be conducted by varying various process parameters and other new profiles.
- It also increase maximum possibility for fabrication of composites by pentagonal pool pin profile than mostly used circular tool pin profile because pentagonal tool pin profile gives fine grained microstructure with any cracks and cavities in processing (nugget) zone.
- Best interaction can be characterized between tool profile material and work piece material.

REFERENCES

- Aydın H, Bayram A, Uguz A, Akay KS (2009),—*Tensile properties of friction stir welded joints of 2024 aluminium alloys in different heat-treated-state"*. vol 38, Mater Des, pp: 2211–2221
- Aldajah SH, Ajayi, Fenskeb GR, Davidc S. (2009), *"Effect of friction stir processing on the tribological performance of high carbon steel"*. vol 5, journal of Wear 2009, pp: 350–355.
- B.C. Liechty, B.W. Webb (2008), *"Flow field characterization of friction stir processing using a particle-grid method"*.vol 8, journal of materials processing technology 208, pp: 431–443.
- B. Zahmatkesh, M.H. Enayati (2010),*"A novel approach for development of surface nanocomposite by friction stir processing"* vol 33,Materials Science & Engineering A,pp:125.
- Buffa G, Fratini L,Pasta S, Shivpuri R (2008),—*On the thermo-mechanical loads and the resultant residual stresses in friction stir processing operations"*. Volume 57, CIRP Annals - Manufacturing Technology, pp: 287–290.
- C.J. Hsu, P.W. Kao (2006),—*Intermetallic-reinforced aluminium matrix composites produced in situ by friction stir processing"*.vol. 55, Materials Letters, pp: 1315–1318.
- Cavaliere P. (2005),—*Mechanical properties of friction stir processed 2618/Al2O3/20 metal matrix composite*‖. Vol. 33,Materials Science & Engineering Composites: pp:1657–1665.
- Cavaliere. P, Squillace.A (2005),*"High temperature deformation of friction stir processed 7075 aluminium alloy"*. vol 55, Materials Characterization (2005).pp: 136– 142.
- Charit I, Mishra R.S (2003), —*High strain rate super plasticity in a commercial 2024 Al alloy via friction stir processing"*.vol 9, Mater Eng A pp: 290–296.
- Charit I, Mishra RS(2008),*"Abnormal grain growth in friction stir processed alloys"*.vol 56,Scr Mater,pp:367–371.
- Douglas C. Kenneth S. Vecchio (2007), —*Thermal history analysis of friction stir processed and submerged friction stir processed aluminium"*.vol 465, Materials Science and Engineering A ,pp: 165–175.
- Darras B.M,Khraisheh M.K,Abu-Farha F.K,OmarM.A(2007),*"Friction stir processing of commercial AZ31 magnesium alloy"*. vol 191, Journal of Materials Processing Technology 191 (2007), pp: 77–81.
- Essam R.I. Mahmouda,b, Makoto Takahashib, Toshiya Shibayanagib, Kenji Ikeuchib(15 January 2010), *"Wear characteristics of surface-hybrid-MMCs layer fabricated on aluminium plate by friction stir processing"*. vol 268,Wear journals 268 ,pp: 1111–1121.

- Essam RIM, Takahashi T, Shibayanagi T, Ikeuchi K.(2005), —*Wear characteristics of surface-hybrid-MMCs layer fabricated on aluminium plate by friction stir processing".*vol 5, Wear journal pp: 1111–1121.
- Hassan kaa, Norman AF, Price da, Prangnell PB (2003). —*Stability of nugget zone grain structures in high strength Al alloy friction stir welds during solution treatment".* vol 6,Acta Mater, pp: 1923–1936.
- Hsu c.j, chang c.y, kao k.w, ho n.j, chang c.P.(14 September 2006), *"Al–Al3Ti nanocomposites produced in situ by friction stir processing".*vol 54, Acta Material.pp: 5241–5249.
- Hsu C.J (24 July 2006)*," Intermetallic-reinforced aluminium matrix composites produced in situ by friction stir processing".* Vol 61,Materials Letters 61 (2007) pp: 1315–1318.
- Jerome S, Govind Bhalchandra S. Kumaresh S.P, Babu, Ravisankar B(2007)," *Influence of Microstructure and Experimental Parameters on Mechanical and Wear Properties of Al-TiC Surface Composite by FSP Route"* Vol. 11.Journal of Minerals & Materials Characterization & Engineering, No.5, pp.493-507.
- Johannes L.B,Charit I, Mishra R.S,Ravi Verma(5 February 2007),*"Enhanced superplasticity through friction stir processing in continuous cast AA5083 aluminium".* Vol 23, Materials Science and Engineering A ,no 464,pp: 351–357.
- Karthikeyan L, Senthilkumar VS, Balasubramanian V, Natarajan S (2009),—*Mechanical propey and microstructural changes during friction stir processing of cast aluminium 2285 alloy".* Vol 12, Material Des paper; pp: 2237–2242.
- Karthikeyan(7 AUG 2009),*"On the role of process variables in the friction stir processing of cast aluminium A319 alloy".* Materials and Design. pp: 761–771.
- Kok M (2006), —*Abrasive wear of Al2O3 particle reinforced 2024 aluminium alloy composites fabricated by vortex method".* Composites: Part A, pp: 457–464.
- Liu F.C, Xiao B.L,Wang K, Ma Z.Y.(16 March 2010)." *Investigation of superplasticity in friction stir processed 2219Al alloy".*vol 13, Materials Science and Engineering A 527, pp: 4191–4196.
- Liming Ke (2010), —*Al–Ni intermetallic composites produced in situ by friction stir processing".* vol 55,Journal of Alloys and Compounds,pp: 494–499.
- Mishra R.S., Z.Y. Ma, I. Charit (2002), —*Friction stir processing: a novel technique for fabrication of surface".* A review journal Materials Science and Engineering A341,pp: 307- 310.
- Mishra R.S, MA Z.Y (2005), —*Friction stir welding and processing".* Materials Science and Engineering, pp: 1–78.

- Mondal A.K, Kumar S.(24 December 2008), *"Dry sliding wear behaviour of magnesium alloy based hybrid composites in the longitudinal direction"*. Wear 267,pp: 458–466.
- Ming Zhou, Henry Hu, Naiyi Li, and Jason Lo (May 20, 2005), *"Microstructure and Tensile Propertiesof Squeeze Cast Magnesium Alloy AM50"*.vol 14,JMEPEG (2005) 14:pp:539-545
- Mishra RS, Mahoney MW (2002), —*Superplastic deformation behaviour of friction stir processed 7075Al alloy*. Acta Mater; pp: 4419–4430.
- McNelley TR, Swaminathan S, Su JQ(2008), —*Recrystallization mechanisms during friction stir welding processing of aluminium alloys*. Scr Mater,pp:349–354.
- Mishra RS, Ma ZY(2005), *"Friction stir welding and processing"*.vol 2, A review journal Mater Sci Eng R, pp: 1–78.
- Olivier Lorraina, Véronique Favierb, Hamid Zahrounic, Didier Lawrjaniecd (11 November 2009), *"Understanding the material flow path of friction stir welding process using unthreaded tools"*.vol 11, Journal of Materials Processing Technology 210,pp: 603–609.
- Poggie RA, Wert JJ(1992), —*The role of oxidation in the friction and wear behavior of solid solution Cu–Al alloys in reciprocating sliding contact with sapphire and D2 tool steel"*. Wear paper ,pp:315–326.
- Srinivason Swaminathan *et al.* (2009), —*Peak Stir Zone Temperatures during Friction Stir Processing"*. Minerals, Metals & Materials Society and ASM International page no: 102.
- T.R. mcnelley.(26 November 2007),—*Recrystallization mechanisms during friction stir welding/processing of aluminium alloys"*. Vol 8,Scripta Materialia 58,pp: 349 354
- Widener CA.(2005), —*Evaluation of post-weld heat treatments for corrosion protection in 2024 and 7075 aluminium alloys"*. Ph.D. Thesis, Wichita State University, College of Engineering, Dept. of Mechanical Engineering; 2005.
- Yazdipoura, A. Shafiei Mc, K. Dehghanib(19 August 2009*)*, *"Modeling the microstructural evolution and effect of cooling rate on the nanograins formed during the friction stir processing of Al5083"*. vol 52,Materials Science and Engineering A 527 .pp: 192–197
- Y. Morisada et al (11 November 2008),*"Nanostructured tool steel fabricated by combination of laser melting and friction stir processing"*.vol 505, Materials Science and Engineering A.pp: 157–162
- Zahmatkesh B, Enayati M.H, Karimzadeh F (2010),*"Tribological and microstructural evaluation of friction stir processed Al2024 alloy"*. vol 505,Materials and Design.pp:57–62